Android™ Tablet
Application Development

FOR

DUMMIES®

Android™ Tablet
Application Development

FOR
DUMMIES®

by Donn Felker

WILEY

John Wiley & Sons, Inc.

Android™ Tablet Application Development For Dummies®

Published by
John Wiley & Sons, Inc.
111 River Street
Hoboken, NJ 07030-5774

www.wiley.com

Copyright © 2012 by John Wiley & Sons, Inc., Hoboken, New Jersey

Published by John Wiley & Sons, Inc., Hoboken, New Jersey

Published simultaneously in Canada

WILEY

About the Author

Donn Felker is a recognized leader in the development of state-of-the-art, cutting-edge software for the Web and for mobile devices. He is an independent consultant with over 11 years of professional experience in various markets, including entertainment, health, retail, insurance, financial, and real estate. He is a mobile junkie, serial entrepreneur, and creative innovator in all things mobile and Web. He is the CTO of QONQR, a geo-social game of world domination. He is a partner in AgileMedicine, a company that uses emerging technology to advance healthcare and medical research. He is the founder of Agilevent, an innovative creative development firm that has done work for small startups as well as for Fortune 500 companies. He is the author of *Android Application Development For Dummies*. He is a Microsoft ASP Insider, an MCTS for .NET Framework 2.0 and 3.5 Web Applications, and a certified ScrumMaster. He's a national speaker on topics that include Android, .NET, and software architecture. He is a writer, presenter, and consultant on various topics ranging from architecture to development in general, agile practices, and patterns and practices. Follow Donn on Twitter (@donnfelker) or read his blog here: http://blog.donnfelker.com.

Dedication

To my son, Michael, whose smile and baby jibber jabber would crack me up each time he looked at me. Thanks son, I love you.

To my gorgeous daughter, Sophia, who has proved that a quick batting of her eyes can win over this old man in almost any situation. I love you.

To my beautiful and caring wife, Ginamarie, who survived three brutal winters in Minneapolis so I could experience living in a place I always wanted. Thank you for giving me two gorgeous children and for being so very patient and supportive of my insane workload over the last year. I love you. Time for some vacations! Let's go!

Last, but certainly not least . . . to Dad. Thank you for reconnecting and letting the truth be known. It means more to me than you will ever know. I love you.

Author's Acknowledgments

Thanks to Wiley acquisitions editor Kyle Looper for giving me more than a few fair extensions when writing this book. I really appreciate the help, support, and insight into everything publishing-related. Thank you for your patience and everything else.

Thanks to editor Christopher Morris for being patient and a diligent editor.

Copy editors Beth Taylor and Teresa Artman helped find some of my most subtle mistakes, which allowed me to polish the book content even more. Thank you.

Technical editor Andre Taddeini is one of the few technical individuals I trust reviewing my work. You're my friend, and I'm glad you could help me out again. Your "sanity check" of the technical accuracy of the book was outstanding. Without you, this book would have been a mess. Thank you.

Thanks to Michael Burton whose knowledge of Android (and of development in general) blows my mind each time we work together. Thank you for the metric ton of great insight and guidance that you provided in regards to Java and the Android framework. Had I not worked beside you I would have missed a lot of great info. Thank you!

Publisher's Acknowledgments

We're proud of this book; please send us your comments at http://dummies.custhelp.com.
For other comments, please contact our Customer Care Department within the U.S. at 877-762-2974,
outside the U.S. at 317-572-3993, or fax 317-572-4002.

Some of the people who helped bring this book to market include the following:

Acquisitions, Editorial, and Vertical Websites

Sr. Project Editor: Christopher Morris

Acquisitions Editor: Kyle Looper

Copy Editors: Beth Taylor, Teresa Artman

Technical Editor: Andre Taddeini

Editorial Manager: Kevin Kirschner

Vertical Websites Project Manager:
Laura Moss-Hollister

Vertical Websites Assistant Project Manager:
Jenny Swisher

Vertical Websites Associate Producers: Josh
Frank, Marilyn Hummel, Douglas Kuhn,
and Shawn Patrick

Editorial Assistant: Amanda Graham

Sr. Editorial Assistant: Cherie Case

Cover Photos: ©istockphoto.com/
Dean Turner; ©istockphoto.com/
Linda Bucklin

Cartoons:
Rich Tennant (www.the5thwave.com)

Composition Services

Project Coordinator: Sheree Montgomery

Layout and Graphics: Carl Byers,
Joyce Haughey

Proofreaders: Laura Albert, Laura Bowman,
Melissa Cossell

Indexer: Rebecca Salerno

Publishing and Editorial for Technology Dummies

Richard Swadley, Vice President and Executive Group Publisher

Andy Cummings, Vice President and Publisher

Mary Bednarek, Executive Acquisitions Director

Mary C. Corder, Editorial Director

Publishing for Consumer Dummies

Kathleen Nebenhaus, Vice President and Executive Publisher

Composition Services

Debbie Stailey, Director of Composition Services

Contents at a Glance

Table of Contents

Part II: Building and Publishing Your First Android Tablet Application 93

Introduction

Welcome to *Android Tablet Application Development For Dummies,* the second *For Dummies* book that covers Android application development. You can consider this the second installment of the Android series. The first *(Android Application Development For Dummies)* covers phone development, and this book covers tablet development. I hope you enjoy finding out about how to program for the Android platform from this book as much as I enjoyed writing it!

When Android was acquired by Google in 2005 (yes, Android was at one point a start-up company), I'll be honest, I didn't have much interest in it. I heard that Google might be entering the mobile space, but as with anything in the technology industry, I didn't believe it until I saw it firsthand. Fast-forward to a few years later, when Google announced its first Android phone: the G1. When I heard this news, I was glued to the computer, reading reviews, watching videos, and researching the product as much as I could. I knew that this product would be the start of something huge.

I got my start in Android development about a week after my wife received her G1, the first publicly released Android device. At that time, the G1 didn't offer the rich feature set of the iPhone, but I believed in the platform. As soon as Donut (Android 1.6) was released, it was evident that Google was putting some effort into the product.

Today, we're on version 3.2 of the Android platform, with the next version of Android (codename *Ice Cream Sandwich*) just around the corner. The platform is barely three years old, and I see no sign of its development slowing down. Without doubt, this is an exciting time in Android development — an excitement that should be infectious as you read this book and begin to work on your own applications.

About This Book

Android Tablet Application Development For Dummies is a beginner's guide to developing Android applications.

The Android platform is a *device-independent* platform, which means that you can develop applications for various devices. (These devices include but aren't limited to phones, e-book readers, netbooks, and GPS devices. Soon, television sets will join the list. Yes, you read it correctly — TV! Google has announced plans to include a Google TV offering in the Android platform.) However, this book emphasizes development for tablet devices above all others.

Finding out how to develop for the Android platform opens a large variety of development options for you. This book distills hundreds, if not thousands, of pages of Android documentation, tips, tricks, and tutorials into a short, digestible format that allows you to springboard into your future as an Android developer. This book isn't a recipe book, but it gives you the basic knowledge to assemble various pieces of the Android framework to create interactive and compelling applications.

Conventions Used in This Book

Throughout the book, you use the Android framework classes, and you will be creating Java classes and XML files.

Code examples in this book appear in a monospace font so that they stand out from other text in the book. This means that the code you'll see looks like this:

```
public class MainActivity
```

Java is a high-level programming language that is case-sensitive, so be sure to enter the text into the editor *exactly* as you see it in the book. I also use the standard Java conventions in this book. Therefore, you can transition easily between my examples and the example code provided by the Android Software Development Kit (SDK). All class names, for example, appear in `PascalCase` format, and all class-scoped variables start with `m`.

Foolish Assumptions

To begin programming with Android, you need a computer that runs one of the following operating systems:

- ✔ Windows XP (32 bit), Vista (32 or 64 bit), or Windows 7 (32 or 64 bit)
- ✔ Mac OS X (Intel) 10.5.8 (x86 only)
- ✔ Linux (i386)

You also need to download the Android SDK (which is free) and the Java Development Kit (or JDK, which is also free), if you don't already have them on your computer. I explain the entire installation process for all the tools and frameworks in Chapter 3.

You don't need any Android application development experience under your belt to get started. I expect you to approach this material as a blank slate because the Android platform accomplishes various mechanisms by using different paradigms that most programmers aren't used to using — or developing with — on a day-to-day basis. However, because Android applications are developed in the Java programming language, I expect you to be familiar with that language. You don't have to be a Java guru, but you should understand the syntax, basic data structures, and language constructs. Also, because XML is also used to define various resources inside Android applications, I advise you to have some understanding of that language before you begin. I don't expect you to be an expert in these languages, however. I started in Android with a background in C#, having worked only with Java in college nearly ten years earlier, and I fared just fine.

You don't need a physical Android device, because all the applications you build in this book work on the emulator. I highly recommend developing on a real device, however, because it allows you to interact with your applications as real users would.

How This Book Is Organized

Android Tablet Application Development For Dummies has four parts, which I describe in the following sections.

Part 1: The Nuts and Bolts of Android Tablets

Part I introduces the tools and frameworks that you use to develop Android applications. It also introduces the various SDK components and shows you how they're used in the Android ecosystem.

Part 11: Building and Publishing Your First Android Tablet Application

Part II introduces you to building your first Android application: the Screen Brightness Toggle application. After you build the initial application, I show you how to create an app widget for the application that you can place on the home screen of the Android device. I tie everything together by demonstrating how to publish your application to the Android Market.

Part III: Creating a Feature-Rich Application

Part III takes your development skills up a notch by walking you through the construction of the Task Reminder application, which allows users to create various tasks with reminders. I cover the implementation of an SQLite database in this multiscreen application. You also see how to use the Android status bar to create notifications that can help increase the usability of your application.

Part IV: The Part of Tens

Part IV brings together the prizes that I've found through my trials and tribulations in Android development. I give you a tour of sample applications that prove to be stellar launching pads for your Android apps, and I introduce useful Android libraries that can make your Android development career a lot easier.

Icons Used in This Book

This icon indicates a useful pointer that you shouldn't skip.

This icon represents a friendly reminder about a vital point you should keep in mind while proceeding through a particular section of the chapter.

This icon signifies that the accompanying explanation may be informative but isn't essential to understanding Android application development. Feel free to skip these snippets, if you like.

This icon alerts you to potential problems that you may encounter along the way. Read and remember these tidbits to avoid possible trouble.

Where to Go from Here

It's time to explore the Android platform! If you're a bit nervous, let me assure you that you don't have to worry; you should be nervous only because you're excited.

If you're ever unsure about anything in the code, you can download the full source code from my GitHub account, located at `http://github.com/donnfelker`. From time to time, I provide code updates to the source. You can also find other examples in my other source repositories stored on the same site. Finally, you can find the same material on the *For Dummies* website at `www.dummies.com/go/androidtabletappdevfd`.

Part I

The Nuts and Bolts of Android Tablets

The 5th Wave   By Rich Tennant

"It's an Android wedding planner. It'll create all your checklists and timetables, and after the ceremony it turns all the documents into confetti and throws it in your face."

In this part . . .

Part I introduces you to the Android platform and describes what makes a spectacular Android application. I briefly explore various parts of the Android software development kit (SDK) and explain how you can use them in your applications. I also guide you through the process of installing the tools and frameworks necessary to develop Android applications.

Chapter 1

Developing Spectacular Android Tablet Applications

Google's Android platform has taken off. Since the first *Android Application Development For Dummies* book was released, Android has gained a tremendous amount of traction in regard to market share in the smartphone space. When first released, Android was considered the underdog of the mobile field but is now considered one of the juggernauts. Google acquired the Android project in 2005 (see the sidebar "The roots of Android" later in this chapter) to ensure that a mobile operating system (OS) could be created and maintained in an open platform. Google continues to pump time and resources into the Android project, which has already proved to be beneficial. As of January 2011, Android reported to have 53 percent of the market share, which proves that Android has gone from a blip on the proverbial mobile radar to a major provider for most mobile consumers. In less than three years, Android has gone from zero to hero in the mobile world and its only going to get larger with each new Android tablet that gets released!

Tablet applications are the talk of the town, and companies around the world cannot wait to release their next groundbreaking application. Newspapers have recently been replaced with tablet applications and subscription-based models, and now more people buy digital books to read on tablet devices than they do physical copies. Enveloping yourself into the world of tablet application development will give you the tools necessary to compete in the ever-growing field of mobile and tablet computing. Tablet development is similar to phone development, but different in a very important way — the form factor. The way that users interact with tablets is much different than the way they interact with phones. Users tend to use tablets for longer periods of time. Tablet users tend to be more cutting-edge, too — having already ridden the phone wave, they've now moved onto the next latest and greatest thing.

Why Develop for Android Tablets?

If you're already developing apps for Android mobile devices, developing for Android tablets is the next logical venture for you. If you're not developing for Android mobile devices, the real question should be, "Why *not* develop for Android tablets?" Do you want your app to be available to millions of users worldwide? Do you want to publish apps as soon as you're done writing and testing them? Do you like developing on open platforms? Do you have a great app in mind more suited for a tablet than a smart phone? If you answered yes to any of these questions, I think you have your answer, but in case you're still undecided, keep reading, and I'll explain what I mean.

Using your existing Android code

If you're an existing Android developer, pay attention: You can use the same code you wrote for your other Android apps on the Android tablet. Yes, that's right, you can develop for both devices with the same code. With Android, you can have your cake and eat it, too! The only differences you'll encounter are some nuances between device screen sizes and how to handle the various new features in both platforms. Android has always aimed itself as being as backward-compatible as possible, and if you refrain from using newer features, you can write apps that work on previous versions. I cover these differences throughout the course of this book.

Major market share

As a developer, you have an opportunity to develop apps for a booming market that's expanding on a daily basis. As of this writing, Android is set to outpace many other carriers in market share in coming months. With so many users, it's never been easier to write an application that can be downloaded and used by real people. And the Android Market puts your app right into your users' hands with little fuss: Users don't have to go searching the

Internet to find an app to install. They just simply go to the Android Market, and they have access to all *your* apps. Because the Android Market comes preinstalled on most Android devices (I discuss a few exceptions later), users typically search the Android Market for all their app needs. Your app's number of downloads can soar in just a few days.

Quick time to market

With all the application programming interfaces (APIs, which are the entry point into the features of the Android framework) that Android comes packed with, developing full-featured applications in a relatively short time frame is easy. After you sign up with the Android Market, just upload your apps and publish them immediately. "Wait," you may ask, "are you sure?" Why, yes, I am! Unlike other mobile marketplaces, the Android Market has no app-approval process. All you have to do is write apps and publish them.

 Technically, anyone can publish anything, but it's good karma to follow Google's Terms of Service and keep your apps family-friendly. Remember that Android users come from diverse areas of the world and from all age categories. Android users can rate and report your application as offensive, so be careful when developing and designing your apps. At the end of the day, if you develop good quality apps, you'll end up with nothing but happy users (except for those few curmudgeons who always seem to sneak into the party).

For more information on publishing your apps to the Android Market, see Chapter 10.

Open platform

The Android operating system is *open platform,* meaning that it's not tied to one hardware manufacturer or one provider. As you can imagine, the openness of Android has allowed it to gain market share quickly. All hardware manufacturers and providers can make and sell Android devices. The Android source code is available at `http://source.android.com` for you to view or modify. Nothing prevents you from digging into the source code to see how a certain task is handled. The open-source code allows phone manufacturers to create custom user interfaces (UIs) and add built-in features to some devices. Because everyone can access the raw Android source code, all developers have an even playing field.

The roots of Android

Most people don't know this, but Google didn't start the Android project. The initial Android operating system was created by a small start-up company in Silicon Valley known as Android, Inc., which was purchased by Google in Summer 2005. The founders of Android, Inc., came from various Internet technology companies such as Danger, Wildfire Communications, T-Mobile, and WebTV. Google brought them onto the Google team to help create what is now the full-fledged Android mobile operating system. One of the founders of Android, Andy Rubin, is the former CEO of both Danger Inc. and Android. He is now the senior vice president of mobile at Google where he oversees the development of the Android platform.

Device compatibility

The Android operating system can run on many devices with different screen sizes and resolutions. Additionally, Android comes with tools to help you develop applications that run on a variety of devices, screen sizes, and resolutions. And Google allows your apps to run only on compatible devices: If your app requires a front-facing camera, for example, only devices with a front-facing camera will be able to see your app in the Android Market. This arrangement is known as *feature detection*.

Compatibility ensures that your apps can run on all devices. For Android devices to be certified as compatible (devices have to be compatible to have access to the Android Market), they must follow certain hardware and software guidelines. The requirements for certified compatible devices change often and are available online for your review at the Compatibility Program Overview page at `http://source.android.com/compatibility/index.html`.

Exploiting mashup capability

A *mashup* combines two or more services to create an application. You can create a mashup by using the camera and Android's location services, for example, to take a picture with the exact location displayed on the image. Or, use a maps API with the contact list to show all your contacts on a map. (See "Integrate Google APIs," later in this chapter.)

You can easily make a ton of apps by combining services or libraries in new and exciting ways. And with all the APIs that Android includes, it's easy to use two or more of these features to make your own app.

Here are a few other mashups to get your brain juices pumping. All this stuff is included for you to use, and it's completely legal and free.

- ✔ **Geolocation and social networking:** Social networking is the "in" thing right now. Suppose you want to write an app that tweets your current location every 10 minutes throughout the day. Use Android's location services and a third-party Twitter API (such as jTwitter), a timer in Java, and you can do just that.

- ✔ **Geolocation and gaming:** Location-based gaming, a great way to really put your users into the game, is gaining popularity. A game might run a background service to check your current location and compare it with the locations of other users of the game in the same area. If another user is within a mile of you, for example, you could be notified, and you could challenge her to take control of that given region. None of this would be possible without a strong platform, such as Android and GPS technology.

- ✔ **Contacts and Internet:** With all these cool APIs at your disposal, it's easy to make full-featured apps by combining the functionality of two or more APIs. You can combine contacts and the Internet to create a greeting card app, for example. Or you may just want to add an easy way for your users to contact you from an app or enable users to send the app to their friends. This is all possible with the built-in APIs.

The sky is the limit. All this cool functionality is literally in the palm of your hand. If you want to develop an app that records the geographic location of the device, you can with ease. Android really opens the possibilities by allowing you to tap into these features easily. It's up to you, as the developer, to put them together in a way that can benefit your users.

Developers can do just about anything they want with Android, so be careful. Use your best judgment when creating and publishing apps for mass consumption. Just because you want live-action wallpaper that shows you doing the hula in your birthday suit doesn't mean that anyone else wants to see it.

Also, keep privacy laws in mind before you harvest your users' contact info for your own marketing scheme.

Java: The Android Programming Language

You don't have to be a member of Mensa to program Android applications. I'm glad, because otherwise, I wouldn't be writing them! Programming for

Android is simple because the default programming language of Android is Java. Although writing Android applications is fairly easy, programming in itself can be a difficult task to conquer.

If you've never programmed before, this book may not be the best place to start. I advise that you pick up a copy of *Beginning Programming with Java For Dummies,* by Barry Burd (Wiley), to learn the ropes. After you have a basic understanding of Java under your belt, you should be ready to tackle this book.

Although the majority of Android is Java, small parts of the framework aren't. Android also encompasses the XML language as well as basic Apache Ant scripting for build processes. I advise you to have a basic understanding of XML before delving into this book.

If you need an introduction to XML, check out *XML For Dummies,* by Lucinda Dykes and Ed Tittel (Wiley).

If you already know Java and XML, congratulations — you're ahead of the curve!

Hardware Tools

Google exposes a plethora of functionality in Android, thus giving developers (even the independent guys) the tools needed to create top-notch, full-featured mobile apps. Google has gone above and beyond by making it simple to tap into and make use of all the devices' available hardware.

To create a spectacular Android app, you should take advantage of all that the hardware has to offer. Don't get me wrong: If you have an idea for an app that doesn't need hardware assistance, that's okay too.

Android devices come with several hardware features that you can use to build your apps, as shown in Table 1-1.

Table 1-1	Android Device Hardware
To Get This Functionality. . .	*. . . Use This Hardware Feature*
Determine the location of the tablet/device	GPS radio
Determine the direction the tablet/device is moving	Built-in compass

To Get This Functionality...	. . . Use This Hardware Feature
Determine whether the tablet/device is facing up or down	Proximity sensor
Determine whether the tablet/device is moving	Accelerometer
Use Bluetooth headphones	Bluetooth radio
Record video	Camera
Read NFC tags	Near Field Communication

Most Android devices are released with the hardware that I discuss in the following sections, but not all devices are created equal. Because Android is free for hardware manufacturers to distribute, it's used in a wide range of devices, including some made by small manufacturers overseas — and it's not uncommon for some of these phones to be missing a feature or two.

Also, as the technology advances, device manufacturers are starting to add features that aren't yet natively supported by Android. Don't worry too much, though, because manufacturers that add hardware usually offer a software development kit (SDK) that lets developers tap into the device's unique feature. For example, as of this writing, only a few devices have come with a front-facing camera. Because these devices are the first of their kind, the device manufacturers have released an SDK that developers can use to access this cool new feature, as well as sample code that lets them implement the feature easily.

Android devices come in all shapes and sizes: phones, tablet computers, and e-book readers. You will find many other implementations of Android in the future, such as Google TV (an Android-powered home appliance) as well as cars with built-in, Android-powered, touchscreen computers. The engineers behind Android provide tools that let you easily deploy apps for multiple screen sizes and resolutions. Indeed, the Android team has done all the hard work for you.

Touchscreen

Android phones have touchscreens, a fact that opens a ton of possibilities and can enhance users' interaction with your apps. Users can swipe, flip, drag, and pinch to zoom by moving a finger or fingers on the touchscreen. You can even use custom gestures for your app, which opens even more possibilities.

Android also supports *multitouch,* which means that the entire screen is touchable by more than one finger at a time.

Hardware buttons are old news. You can place buttons of any shape anywhere on the screen to create the UI that's best suited for your app.

GPS

The Android OS combined with a phone's GPS radio allows developers to access a user's location at any given moment. You can track a user's movement as she changes locations. The social networking app from foursquare is a good example; it uses GPS to determine the phone's location and then accesses the web to determine which establishment or public place the user is in or near.

Another great example is the Maps application's ability to pinpoint your location on a map and provide directions to your destination. Android combined with GPS hardware gives you access to the phone's exact GPS location. Many apps use this functionality to show you where the nearest gas station, coffeehouse, or even restroom is located. You can even use the Maps application's API to pinpoint the user's current location on a map.

Accelerometer

Android comes packed with accelerometer support. An *accelerometer* is a device that measures acceleration. That sounds cool and all, but what can you do with it? Well, if you want to know whether the phone is moving or being shaken, or even the direction in which it's being turned, the accelerometer can tell you.

You may be thinking, "Okay, but why do I care whether the phone is being shaken or turned?" Simple! You can use that input as a way to control your application. You can do simple things like determine whether the phone has been turned upside down and do something when that happens. Maybe you're making a dice game and want to immerse your users in the game play by having them shake the phone to roll the dice. This is the kind of functionality that's setting mobile devices apart from typical desktop personal computers.

SD Card

Android gives you the tools you need to access (save and load) files on the device's *SD Card,* a portable storage medium that you can insert into various phones, tablets, and computers. If a device is equipped with an SD Card, you can use it to store and access files needed by your application. As of Android 2.2, you could install apps on the SD Card. However, if your users have phones that don't get Android 2.2, you're not sunk. Just because some users don't have the option of installing apps on the SD Card doesn't mean that you have to bloat your app with 20 MB of resources and hog the phone's limited built-in memory. You can download some or all of your application's resources from your web host and save them to the phone's SD Card. This makes your users happy and less likely to uninstall your app when space is needed.

Not all devices come with an SD Card installed, although most do. Always make sure that the user has an SD Card installed and that adequate space is available before trying to save files to it, though.

Exploring Android Software Tools

Various Android tools are at your disposal while writing Android applications. In the following sections, I outline some of the most popular tools that you will use in your day-to-day Android development process.

Exploit the Internet

Thanks to the Internet capabilities of Android devices, real-time information is easy to obtain. As a user, you can use the Internet to see what time the next movie starts or when the next commuter train arrives. As a developer, you can use the Internet in your apps to access real-time, up-to-date data such as weather, news, and sports scores. You can also use the web to store some of your application's assets, which is what Pandora and YouTube do.

Don't stop there. Why not offload some of your application's intense processes to a web server when appropriate? This can save a lot of processing time in some cases and also helps keep your Android app streamlined. This arrangement — client–server computing — is a well-established software architecture in which the client makes a request to a server that is ready and willing to do something. The built-in Maps app is an example of a client accessing map and GPS data from a web server.

Build in audio and video support

The Android OS makes including audio and video in your apps a breeze. Many standard audio and video formats are supported, and including multimedia content in your apps couldn't be any easier. Sound effects, instructional videos, background music, streaming video, and audio from the Internet can all be added to your app with little to no pain. Be as creative as you want to be. The sky is the limit.

Maintain contacts

Your app can access user contacts that are stored on the phone. You can use this feature to display the contacts in a new or different way. Maybe you don't like the built-in Contacts application. With the ability to access the contacts stored on the phone, nothing is stopping you from writing your own. Maybe you write an app that couples the contacts with the GPS system and alerts the user when she is close to one of the contacts' addresses.

Use your imagination, but be responsible. You don't want to use contacts in a malicious way (see the next section).

Provide security

Android allows your apps to do a lot! Imagine if someone released an app that went through the contact list and sent the entire list to a server somewhere for malicious purposes. This is why most of the functions that modify the user's device or access its protected content need to have permissions to work. Suppose that you want to download an image from the web and save it to the SD Card. To do so, you need to get permission to use the Internet so that you can download the file. You also need permission to save files to the SD Card. Upon installation of the application, the user is notified of the permissions that your app is requesting. At that point, the user can decide whether he wants to proceed with the installation. Asking for permission is as easy as implementing one line of code in your application's manifest file.

Integrate Google APIs

The Android OS isn't limited to making phone calls, organizing contacts, or installing apps. You have much more power at your fingertips. As a developer, you can use external libraries and integrate maps into your application. To do so, you have to use the Maps APIs that are provided by Google that contain the map widgets.

The KISS principle

It's easy to overthink and overcomplicate things when developing applications. The hardest part is to remember the KISS (Keep It Simple, Stupid) principle. One way to overly complicate your code is to just dive in without understanding all the built-in APIs and knowing what they do. You can go that route, but doing so may take more time than just glossing over the Android documentation. You don't have to memorize it, but do yourself a favor and take a look at the documentation. You'll be glad you did when you see how easy it is to use the built-in functionality and how much time it can save you. You can easily write multiple lines of code to do something that takes only one line. Changing the volume of the media player or creating a menu is a simple process, but if you don't know the APIs, you may end up rewriting them and in the end causing yourself problems.

When I started with my first app, I just dived in and wrote a bunch of code that managed the media player's volume. If I'd just looked into the Android documentation a little more, I'd have known that I could handle this with one line of code that's strategically placed inside my application. The same thing goes for the menu. I wrote a lot of code to create a menu, and if I'd only known that a menu framework already existed, it would have saved me several hours.

Another way to really muck things up is to add functionality that isn't needed. Most users want the easiest way to do things, so don't go making some fancy custom tab layout when a couple of menu items will suffice. Android comes with enough built-in controls (widgets) that you can use to accomplish just about anything. Using the built-in controls makes your app that much easier for your users to figure out because they already know and love these controls.

Pinpointing locations on a map

Say you want to write an app that displays your current location to your friends. You could spend hundreds of hours developing a mapping system, or you could just use the free Google Android Maps API.

You can embed and use the API in your application to show your friends where you are, saving you hundreds of hours and cents. Imagine all the juicy map goodness with none of the work developing it. Using the Maps API, you can find just about anything with an address; the possibilities are endless. Display your friend's location, the nearest grocery store, or the nearest gas station — anything or anyplace with an address.

And the Android Maps API can also access the Google Navigation API. Now you can pinpoint your location and also show your users how to get to that location.

Messaging in the clouds

The Android Cloud to Device Messaging framework allows you to send a notification from your web server to your app. Suppose that you store your application's data in the cloud and download all the assets the first time your app runs. But what if you then realize, after the fact, that one of the images is incorrect? For the app to update the image, it needs to know that the image changed. You can send a cloud-to-device message (a message from the web to the device) to your app, letting it know that it needs to update the image. This works even if your app is not running. When the device receives the message, it dispatches a message to start your app so that it can take the appropriate action.

Chapter 2

Switching to Tablet App Development

In This Chapter

▶ Juxtaposing the tablet and phone experience

▶ Developing for the tablet form factor

▶ Expanding tablets for the masses

*O*ur phones, tablets, and other little pieces of metal and plastic consume our day-to-day free time. What's really the difference and why does it matter to you, the developer? Well, for you, the difference matters in some instances but not in others. And although the code you write may be the same, the user interface (UI) is completely different, and the usability of both of these platforms can differ quite a bit. In this chapter, I delve into the differences you'll encounter (and need to account for) while developing for tablet devices using the Android 3.2 platform.

Seeing How Tablets Have Evolved

The term "tablet" isn't new: It's been around for decades. Unfortunately, not one manufacturer and operating system (OS) provider has nailed the implementation enough to make a large splash in the computing market. Tablets were introduced as smaller computers with computer OSes, such as Windows XP, Vista, and 7. These devices resembled laptops, but the screens sometimes swiveled, allowing them to fold backward and sit atop the keyboard, simulating a tablet that you'd write on.

At first, most tablets were controlled via a physical pen interface. Users eventually wanted the ability to use these tablets as computers and usually just opened up the keyboard and performed the functions they needed to perform with the keyboard open. In the end, the user didn't use the tablet as it was designed (for the most part). Implementations changed, designs changed, and new innovations occurred in this space. Unfortunately, though, none of these implementations caught on, mainly becauseusers wanted to use these devices as laptops, not tablets.

Fast-forward to June 2007 when the first iPhone was released and a new generation of phone was born. Android quickly came to market and began to dominate market share. Phone devices have consumed the mobile market and just recently the same market has been infused with a new contender — the redesigned tablet.

Apple introduced the iPad in early 2010, creating uproar in the mobile device market by introducing a tablet with a form factor that even though it was unproven, it still managed to emerge as a winner (as proven through popularity and sales). The iPad changed the way users interacted with tablets. Most users (myself included) found themselves using the device for things they never thought they'd use it for:

✔ **Reading books**

   I had lost my interest in paper books prior to the new tablets.

✔ **Playing quick fun games**

✔ **Using it as a learning tool for children**

   Many apps are very educational.

✔ **Watching movies**

Android 3.0 was still in development, and the only Android tablets on the horizon were the Samsung Galaxy Tab and the Dell Streak. Unfortunately, these Android tablets were released with the 2.x Android framework. The Android team informed the world that the 2.x version of Android was not built or optimized for a tablet experience and that they were working on a new version. Finally, in February 2011, the first Android 3.0 tablet was released — the Motorola XOOM.

As of this writing, I've had the XOOM tablet for a few months, and I can honestly say that the Android 3.1 operating system is stellar. Having owned an iPad for more than six months, I can also easily say that this is the iPad's greatest competitor and that I enjoy the Android 3.1 experience of the XOOM much more than I do the iPad. While my reasons may be personal, I still do believe that Android 3.1 is a significant contender in the tablet market. And although the iPad and XOOM run on different operating systems, they do share one common mechanic — the form factor.

Understanding the Tablet Form Factor

Is that a burnt toaster pastry you're holding to your ear, or is that a phone? Yeah, you've seem them — the guys holding a huge rectangle phone to their ear. It seems that the device manufacturers are constantly trying to reinvent. Would you do that with a tablet? No, probably not, unless you're watching funny cat videos on YouTube and you can't hear the volume. But this brings up the important point of *form factor.*

Keep that previous image in your head — you holding a tablet held to your ear. That doesn't seem right, now does it? But a phone, sure, why not? That seems plausible. Form factor — in this instance — is the form of the device itself. The size of my Nexus One is 2.3" wide and 4.6" in height. Juxtapose this with the size of the XOOM tablet at 6.6" wide and 9.8" in height (assuming the device is held in portrait position). Just by looking at the dimensions alone, you can see that they are completely different in size. Why does this matter to you? Simple — if the size is different, the user experience is different.

The size of something that you hold in your hand dictates how you interact with it. Imagine how you hold a newspaper — probably with both hands. Now imagine reading a single-page pamphlet — probably with one hand. The same goes for the usage of phones and tablets. Most users hold a tablet with two hands but a phone with one hand; see Figure 2-1.

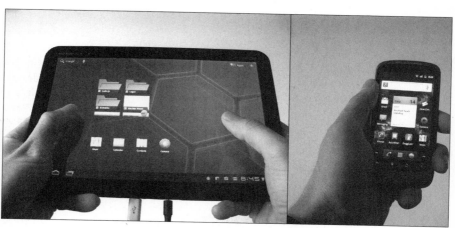

Figure 2-1: Holding a Motorola XOOM with two hands on the left and a Nexus One with one hand on the right.

You can see that the form factor of these devices is completely different. The tablet requires using both hands to operate, but the phone requires only one hand to operate. Developing applications that accommodate to this form factor is the key to success when designing user interfaces for these devices.

User Interface Considerations for Tablets

The Java and Android code that you write to develop Android applications doesn't differ between the two platforms. Both are Java and XML, as you'll see later in this book. However, when designing UIs, you have to take a few scenarios into account:

- ✔ Hand placement
- ✔ Screen size
- ✔ Device (screen) orientation

Hand placement is your number one consideration when developing an application for the tablet market because where a user places his hands on the device dictates how the user interacts with your app. When holding a tablet, users typically hold on to the device on the bottom left and bottom right of the device, as shown in Figure 2-2.

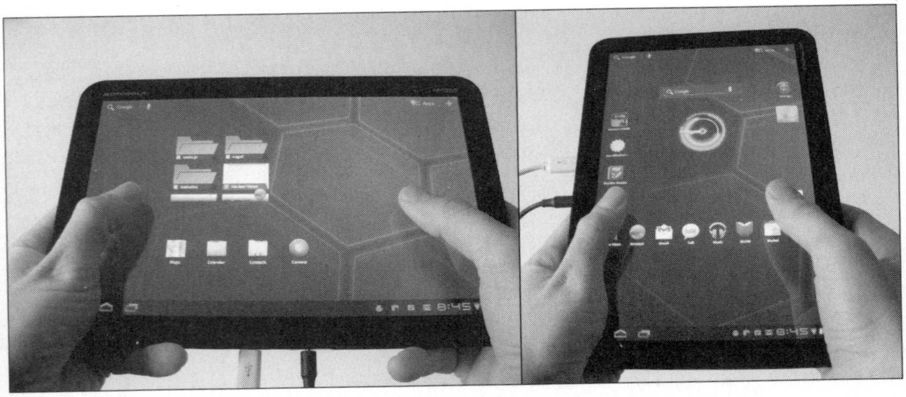

Figure 2-2:
Holding the device in landscape (l) and portrait (r) modes.

So, imagine that you're creating a game, and you need to place game controls on the screen. You don't want to put these controls at the top of the screen on a tablet because the user wouldn't be able to reach them — the game would be more difficult to play from a form factor standpoint.

However, if you were developing this application for a phone (or smaller device), placing controls at the top of the screen would be acceptable because the user could easily move a finger to reach those controls. With the large screens of tablets, doing this just isn't feasible.

So, does this mean that you can't ever place controls at the top of the screen? No, it does not. Just be careful what controls you place at the top of the screen. For example, you wouldn't want to place controls at the top of the screen that are high-frequency usage controls, such as the keyboard or Back button (unless that's on the left side of the screen — where it makes sense for it to be). Instead, those should be placed near the thumbs of the user, located on the bottom right and left of the device (again, depending on the orientation in which the user has the device positioned).

Screen size is another important factor that needs to be taken into consideration when designing the UI of your application. Tablets have much larger screens than their phone counterparts. If your application is a spreadsheet application, I suggest using the large screen of a tablet to emphasize the feature set of your application.

Device manufacturers make Android devices in all shapes and sizes. For example, the Samsun Galaxy Tab has a 7.5" screen. Compared with the XOOM tablet, that's a 3-inch difference in size! When designing the UI, be sure that you lay out your views in a particular manner to allow the design to shrink and grow automatically. I show you how to do this in Chapter 6.

Device orientation also plays a very important factor when designing your application. Some users prefer to hold their device a certain way. Some applications, though, require the user to hold the device in landscape (or portrait) mode in order for it to operate effectively. You can handle all of these orientation changes and layouts through resources and layouts. I'll cover some of the details of how this is implemented in Chapter 5.

Tablets for You, Me, and Everyone Else

As I mention earlier, I have the Motorola XOOM tablet, and I love it. However, some people don't like how big it is and prefer the Samsung Galaxy Tab because it's smaller and more portable. This is the beauty (and downfall) of an open system such as Android. Its beauty is recognized by the fact that we have so many great options for users to consume our applications, but at the same time, this is a downfall. Each device is made with different hardware and is also made with a different screen size. So you can see how certain device sizes and models play directly into your end goal as a developer — developing an application that helps change the world.

Wait, did I really say that? Yes, I did. You sure can develop an application that can change the world. I mean it! Because there are so many different tablets, you can then use these options to play directly into your plans.

Let me explain in more detail. Remember the last time you went to the doctor? He was probably wearing one of those white lab coats — right? — the kind provided by the hospital or medical facility. But have you ever noticed how big the pockets are on those devices? They're actually pretty small. A Motorola XOOM wouldn't fit into the pocket sleeve. However, a Samsung Galaxy Tab does fit perfectly into the sleeve.

Another start-up that I'm part of, Agile Medicine, uses a smaller tablet (such as the Samsung Galaxy Tab) form factor to develop cutting-edge software for the medical industry. Doctors and other medical professionals can use these devices to perform a wide array of operations in a health and medical capacity, meaning that these health-care professionals can become more productive while taking care of you and me on a day-to-day basis. The same thing could not be done for any of the larger tablets.

Specializing in a particular size of tablet can work out for the better if you just think out of the box. For example, you could develop applications for the logistics industry that allow drivers (think UPS/FedEx) to scan, check in and out, and have customers sign for packages all through your app. Other examples include an app that allows auto body shops to provide repair quotes via the tablet. By circling or coloring damaged areas of a vehicle illustration, clients can see what the quote is referencing. Other possibilities include student assistant learning aids, communication devices, and control flow dashboards for manufacturing companies.

Chapter 3

Developing for Android Tablets

Developing for Android tablets is the next wave of application development. The form factor of the tablet allows users of your application to interact with a wider feature set than was possible with a phone application. The larger surface screen is optimal for reading, displaying complex charts and data points, as well as watching movies while on the train on the way home from work. Overall, experiencing an app on a tablet is much smoother, end-to-end, than doing so on a phone.

The Android tablet development process is not much different than developing for the phone, but each development environment does have its own set of gotchas to look out for. The basics of the development process remain much the same. By the end of this chapter, you'll have an application you can publish to the Android Market.

The Android Development Process

The Android development process is quite simple. Using high-level Java language, you can write and create compelling apps that engage your users. Another great thing about Android is that the financial barriers to entry are low. You can start building applications with freely available tools on the Internet. This all means you can focus on writing feature-rich apps instead of figuring out how to just get started.

Installing the tools, frameworks, and SDKs

The tools, frameworks, and SDKs are freely available online, so you can begin writing applications quickly. The frameworks and SDKs are easily installed by following a simple process that is outlined in Chapter 4 as well as online in the Android SDK documentation. Google has done a great job of helping you get started quickly and for as little financial commitment as possible.

Designing your feature-rich application

A feature-rich application could be as simple as allowing the user to swipe left and right to expose new screens, or it could be an e-mail client that notifies you when you have new e-mail by vibrating the device. Designing a feature-rich application does not mean that the graphics are gradients and drop shadows look perfect, rather it means that the application you're designing has the necessary features to go beyond what the end user expected. The best advice I can provide is to think outside the box and use a mind-mapping tool, such as Mind Meister (www.mindmeister.com) to help facilitate your genius app creation.

Coding the application

Thankfully, you don't have to be a professor in computer science to write mobile applications. You can now use the high-level Java programming language to create compelling Android applications. Writing code is simple — you can lean on three simple concepts:

- ✔ **Compiling:** The process that your Java code must go through before your application can be run on the device
- ✔ **Debugging:** The process that allows you to find bugs and walk through the code line by line
- ✔ **Testing:** The process of writing code to test your other code

Packaging your application as an APK

Before you can place your application into the hands of hungry Android Market users, you have to perform one last step — transforming your application into an APK (Android Package File). Your application code is compiled and transformed into (through a build process handled by your compiler) an APK in order for you to distribute your application. Without an APK you'd have tens (if not hundreds or thousands) of files that you'd need to distribute. An APK is a package of files that contains everything necessary to run your application. An APK is the file that the various Android Markets accept in order to distribute your application.

Submitting your app to the Android Market

Submitting your application to the Android Market (and other markets, such as the Amazon App Store) is a simple process. Create an account online, provide a description and a couple screenshots, and upload your APK. That's all there is to it — your app is in the Android Market. I cover submitting your application to the market in Chapter 11.

Using Your Visionary Talent

I've got talents, you've got talents, we've all got one area or another in which we're experts. Perhaps you're an expert home beer brewer and have an idea for an application that allows home brewers to track the brewing process through their device. Also consider other passions and hobbies, and create an application to fit that niche. Using ideas that genuinely interest you is a great way to get into Android development. You'll find that, by using your own visionary talent and ideas, the end result (the application) will be much more polished and refined than if you were writing an application to track your monthly expenses (unless that is your passion).

Sample Code

Sometimes, the best way to find out about a new programming language is to read some existing code! Thankfully, the great minds over at Google have provided a vast array of sample code that you can download with the Android SDK. I cover installing the SDK in Chapter 4. You can find sample code many other places on the web. I provide ten great resources in Chapter 18.

Looking at Android Programming Basics

Android applications are written in Java — not the full-blown Java that J2EE developers are used to, but a subset of Java that is sometimes known as the *Dalvik virtual machine.* This smaller subset of Java excludes classes that don't make sense for mobile devices. If you have any experience in Java, you should be right at home.

It may be a good idea to keep a Java reference book on hand, but if you have any trouble with a Java concept, you can always search Google to find out what you don't understand. Because Java is nothing new, you can find plenty of examples on the web that demonstrate how to do just about anything.

In Java source code, not all libraries are included (such as third-party libraries that assist in list processing or message queue management, and so on). Verify that the package is available to you. If it's not, an alternative that can work for your needs is probably bundled with Android.

Activities

Think of an *activity* as a container for your UI, like a screen. It contains your UI logic as well as the code that runs it. It's kind of like a form, for you Windows programmers out there. Android applications are made up of one or more activities. Your app must contain at least one activity, but an Android application can contain several. I discuss activities in more detail in Chapters 4 and 6.

Intents

Intents make up the core message system that runs Android. An intent is composed of an action that Android needs to perform (View, Edit, Dial, and so on) and data. The action is the general action to be performed when the intent is received, and the data is the data to operate on. The data might be a contact item, for example.

Intents are used to start activities and to communicate among various parts of the Android system. Your application can either broadcast an intent or receive an intent. The best way to think of intents is by placing the abstract term into laymen's terms. Here's a simple example of intents: Your "intent" is to master Android tablet development by reading this book. In creating this intent, your "action" is to "view" the book and read it. You defined an intent by saying what you want to do and how to do it. That's what intents are in Android (well, that's the short version, of course).

When you broadcast an intent, you're sending a message telling Android to make something happen. This intent could tell Android to start a new activity from within your application, or it could start a different application.

Just because you send a message doesn't mean that something will happen automatically. You have to register an *intent receiver* that listens for the intent and then tells Android what to do, whether the task is starting a new activity or starting a different app. If many receivers (applications) can accept a given intent, a chooser can be created to allow the user to pick the app she wants to use. A classic example is long-pressing an image in an image gallery. (*Long-pressing* means clicking something for a long time to bring up a context menu. It's the tablet equivalent of right-clicking.)

By default, various registered receivers handle the image-sharing intents. One is e-mail, and another is the messaging application (among various other installed applications). Because long-pressing sends this kind of image-sharing intent, more than one possible intent receiver can receive it. When this happens, the user is presented with a chooser asking him how to proceed: use e-mail, messaging, or another application? (See Figure 3-1.)

Figure 3-1:
A chooser
on an
Android
mobile
device.

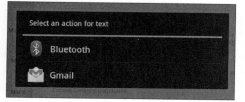

If the Android system cannot find a match for the intent that was sent, and a chooser was not created manually, the application will crash due to a *run-time exception:* an unhandled error in the application. Android expects developers to know what they're doing. If you send an intent that a user's Android device doesn't know how to handle, the device crashes. Best practice is to create choosers for intents that don't target other activities within your application.

Cursorless controls

Unlike PCs, which let you use a mouse to move the cursor across the screen, Android devices let you use your fingers to do just about anything a mouse can do. Of course, that leads to questions like how to program a right-click to bring up a contextual menu? (The short answer is that instead of supporting right-clicking, Android has implemented the long press. Press and hold a button, icon, or screen for an extended period of time, and a context menu appears. So, as a developer, you can create and manipulate context menus.)

Then there's the matter of finding a workaround to using a traditional mouse cursor. Android devices have users use two fingers on an Android device instead of just one mouse cursor, for example. That's cool, but then you have to keep in mind that fingers come in all sizes, and design your UI accordingly. For example, you have to be cognizant to make the buttons large enough, with enough spacing, so that even users with large fingers can interact with your apps easily.

Views and widgets

A *view* is a basic UI element — a rectangular area on the screen that is responsible for drawing and event handling. I like to think of views as being basic controls, such as a label control in HTML. Here are a few examples of views:

- ✔ `ContextMenu`
- ✔ `Menu`
- ✔ `View`
- ✔ `Surface view`

Widgets are more-advanced UI elements, such as check boxes. Think of them as being the controls that your users interact with. Here are a few widgets:

- ✔ `Button`
- ✔ `CheckBox`
- ✔ `DatePicker`
- ✔ `DigitalClock`
- ✔ `Gallery`
- ✔ `FrameLayout`
- ✔ `ImageView`
- ✔ `RelativeLayout`
- ✔ `PopupWindow`

Many more widgets are ready for you to use. For complete details, check out the `android.widget` package in the Android documentation at `http://developer.android.com/reference/android/widget/package-summary.html`.

I cover adding widgets in Chapter 8.

Asynchronous calls

Who called? I don't know anybody named Asynchronous, do you?

You use asynchronous processing for tasks that take a long time — network communication (Internet), media processing, or anything else that might make the user wait. If the user has to wait, you should use an asynchronous call and some type of user interface (UI) element to let him know that something is happening, such as a progress meter.

The `AsyncTask` class in Android allows you to run multiple operations at the same time without having to manage a separate thread yourself. Using `AsyncTask` not only lets you start a new process without having to clean up after yourself, but also returns the result to the activity that started it. This allows you to have a clean programming model for asynchronous processing.

A *thread* is a process that runs separately from and simultaneously with everything else that's happening.

Failing to use an asynchronous programming model can cause users of your application to believe that your application is buggy. Downloading the latest Twitter messages via the Internet takes time, for example. If the network gets slow and you're not using an asynchronous model, the application will lock up, and the user will assume that something is wrong because the application isn't responding to her interactions. If the application doesn't respond within a reasonable time (a time that the Android OS defines), Android presents an Application Not Responding (ANR) dialog box, as shown in Figure 3-2. At that time, the user can decide to wait or to close the application.

Figure 3-2:
An ANR
dialog box.

Best practice is to run CPU-expensive or long-running code inside another thread, as described in the Designing for Responsiveness page on the Android developer site at `http://developer.android.com/guide/practices/design/responsiveness.html`.

Background services

If you're a Windows user, you may already know what a *service* is: an application that runs in the background and doesn't necessarily have a UI. A classic example is an antivirus application that usually runs in the background as a service. Even though you don't see it, you know that it's running.

Most music players that can be downloaded from the Android Market run as background services. This is how you can listen to music while checking your e-mail or performing another task that requires the use of the screen. I show you how to create a background service in Chapter 7.

Making Yourself Seen

During your application's lifetime, you will need to get ahold of your user. Sometimes you need to do this in a way that requires the user to interact with your application before he can continue, and other times you only need to quickly notify a user that something happened (such as "This item is saved" when they click on a Save button). Another option is to inform the user that something has happened by placing an icon in the notification bar for later review. Use this option if the user's attention is not needed immediately.

Sitting at the notification bar

The notification bar is the little area at the bottom-right side of the tablet. This inconspicuous location allows developers to place icons in the bar so that the user can recognize that some type of action needs to take place. For example, the user may be reading an e-mail, and a task reminder application (the one you write later in this book) may need to remind the user that a task needs to be reviewed. You can use the notification bar to place passive action items in front of the user.

Good-looking toast

Toast notifications are small blurbs of information that pop up in front of a user and do not require end-user interaction. A toast notification persists onscreen for a few seconds before it fades away. Here's a good example: A user taps a Save button, and a small message appears informing the user that the item has been saved. The user does not have to see this message (nothing bad would happen if they never saw it), but it may help with the user experience in the end product.

Chapter 4

Prepping Your Development Headquarters

. . . .

In This Chapter

▶ Becoming an Android application developer

▶ Collecting your tools of the trade

▶ Downloading and installing the Android SDK

▶ Getting and configuring Eclipse

▶ Working with the Android ADT

. . . .

All of the basic building blocks you need to develop rich Android applications — the tools, the frameworks, and even the source code — are *free*. No, you don't get a free computer out of it, but you do get to set up your development environment and start developing applications for free, and you can't beat free. Well, maybe you can — such as someone paying you to write an Android application, but keep working at it and you'll get there soon enough.

In this chapter, I walk you through the steps to get and install the necessary tools and frameworks so that you can start building kick-butt Android applications.

Developing the Android Developer Inside You

Becoming an Android developer is a lot simpler than you probably think. To see what's involved, ask yourself the following questions:

✔ Do I want to develop Android applications?

✔ Do I like free software development tools?

✔ Do I prefer to pay no developer fees?

✔ Do I have a computer I can use to develop applications?

If you answered yes to all these questions, today is your lucky day: You're ready to become an Android developer. You may be wondering about the *no fees* part. Yep, you read that correctly: You pay no fees to develop Android applications — that means that the tools, framework, and languages to develop Android applications are 100 percent free.

There's always a catch, right? You can develop for free to your heart's content, but as soon as you want to publish your application to the Android Market — where you upload and publish your apps — you need to pay a nominal registration fee. As of this writing, the fee is $25.

Put your mind at ease about fees. If you're developing an application for a client, you don't necessarily have to pay a fee for client work. You can publish your application as a redistributable package that you can give your client, who can then publish the application to the Android Market by using his Market account. Doing this ensures that you don't have to pay a fee for client work — which means that you can be a bona fide Android developer without ever having to pay a fee. Now, that's cool.

Assembling Your Toolkit

Now that you know you're ready to be an Android developer, grab your computer and get cracking on installing the tools and frameworks necessary to build your first blockbuster application.

Android source code

Be aware that the full Android source code is open source, which means that it's not only free to use, but also free to modify. If you want to download the Android source code and create a new version of Android, you're free to do so. Check the Android Git repository. You can also download the source code at `http://source.android.com`.

Sometimes when you're developing an Android application, you may want to use a resource from the core Android system, an icon for a Settings menu option, for example. By accessing the Android source code, you can browse the various resources and download the resources you need for your project. Having access to the source code also enables you to dig in and see exactly how Android does what it does.

You can visualize the Android source code as a system of discrete layers, each built upon the other. These layers include the following (see Figure 4-1):

- Linux 2.6 kernel
- Android libraries (and Android run time)
- Applications

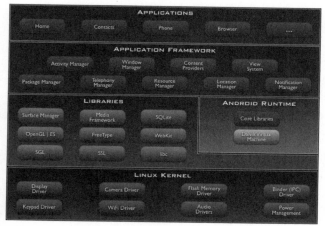

Figure 4-1: The structure of the Android source code.

Linux 2.6 kernel

Android was created on top of the open-source Linux 2.6 kernel. The Android team chose to use this kernel because it provided proven core features on which to develop the Android operating system. The features of the Linux 2.6 kernel include (but aren't limited to) the following:

- **Security model:** The Linux kernel handles security between the application and the system.
- **Memory management:** The kernel handles memory management for you, leaving you free to develop your app.
- **Process management:** The Linux kernel manages processes well, allocating resources to processes as they need them.
- **Network stack:** The Linux kernel also handles network communication.
- **Driver model:** The goal of Linux is to ensure that everything works. Hardware manufacturers can build their drivers into the Linux build.

You can see a good sampling of the Linux 2.6 feature set in Figure 4-1.

Android framework

Atop the Linux 2.6 kernel, the Android framework was developed with various features. These features come from numerous open-source projects. The output of these projects resulted in the following:

- **The Android run time:** The Android run time is composed of Java core libraries and the Dalvik virtual machine.

- **Open GL (graphics library):** This cross-language, cross-platform application programming interface (API) is used to produce 2D and 3D computer graphics.

- **WebKit:** This open-source Web browser engine provides the functionality to display Web content and simplify page loading.

- **SQLite:** This open-source relational database engine is designed to be embedded into devices.

- **Media frameworks:** These libraries enable you to play and record audio and video.

- **Secure Sockets Layer (SSL):** These libraries are responsible for Internet security.

Refer to Figure 4-1 for a list of common Android libraries.

Application framework

You're probably thinking, "Well, that's all nice and well, but how do these libraries affect me as a developer?" It's simple: All these open-source frameworks are available to you through Android. You don't have to worry about how Android interacts with SQLite and the surface manager; you just use them as tools in your Android tool belt. The Android team has built Android on a known set of proven libraries and has given them to you, all exposed through Android interfaces. These interfaces combined the various libraries and made them useful to the Android platform as well as to you as a developer. Android has all these libraries built in the background and exposes these features to you without your having to build any of the functionality that they provide, including the following:

- **Activity manager:** Manages the activity life cycle.

- **Telephony manager:** Provides access to telephony services as well as some subscriber information, such as phone numbers.

- **View system:** Handles the views and layouts that make up your user interface (UI).

- **Location manager:** Finds out the device's geographic location.

Refer to Figure 4-1 to see the libraries that make up the application framework.

Applications

From kernel to application, the Android operating system has been developed with proven open-source technologies. This enables you, as a developer, to build rich applications that have been fostered in the open-source community. (Refer to Figure 4-1.) The Applications section is where your application sits.

Java knowledge

The Java programming language is one of the glorious tools that make programming Android a breeze compared with programming for other mobile platforms. Whereas other languages insist that you manage memory, deallocate and allocate bytes, and then shift bits around like a game of dominoes, Java has a little buddy called the Java Virtual Machine (JVM) that helps take care of that for you. The JVM allows you to focus on writing Java code to solve a business problem by using a clean, understandable programming language (or to build that next, really cool, first-person shooter game you've been dreaming of) instead of focusing on the plumbing just to get the screens to show up.

You're expected to understand the basics of the Java programming language before you write your first Android application. If you're feeling a bit rusty and need a refresher course on Java, you can visit the Java tutorials site at `http://download.oracle.com/javase/tutorial/`.

I cover some Java in this book, but you may want to spend some time with a good book like *Java All-in-One For Dummies,* by Doug Lowe (Wiley), if you don't have any Java experience.

Tuning Up Your Hardware

You can develop Android applications on various operating systems, including Windows, Linux, and Mac OS X. I do the development in this book on a Windows 7 operating system and Mac OS X, but you can develop using Linux instead.

Operating system

Android supports all the following platforms:

- ✔ Windows XP (32-bit), Vista (32- or 64-bit), and 7 (32- or 64-bit)
- ✔ Mac OS X 10.5.8 or later (x86 only)
- ✔ Linux (tested on Ubuntu Linux, Lucid Lynx)

Note that 64-bit distributions must be capable of running 32-bit applications.

Throughout the book the examples use Mac OS X 10.6.6 as the operating system. Therefore, some of the screenshots may look a little different from what you see on your machine. If you're using a Windows machine, your paths may be different. Paths in this book look similar to this:

```
/path/to/file.txt
```

If you're using a Windows machine, however, your paths will look similar to this:

```
c:\path\to\file.txt
```

Computer hardware

Before you start installing the required software, make sure that your computer can run it adequately. I think it's safe to say that just about any desktop or laptop computer manufactured in the past four years will suffice. I wish I could be more exact, but I can't; the hardware requirements for Android simply weren't published when I wrote this book. The slowest computer that I have run Eclipse (the Android development environment) on a laptop with a 1.6-GHz Pentium D processor with 1 GB of RAM. I've run this same configuration under Windows XP and Windows 7, and both operating systems combined with that hardware can run and debug Eclipse applications with no problems.

To ensure that you can install all the tools and frameworks you'll need, make sure that you have enough disk space to accommodate them. The Android developer site has a list of hardware requirements, outlining how much hard drive space each component requires, at http://developer.android.com/sdk/requirements.html.

To save you time, I compiled my own statistics from personal use of the tools and software development kits (SDKs). I found that if you have about 3 GB of free hard-drive space, you can install all the tools and frameworks necessary to develop Android applications.

Installing and Configuring Your Support Tools

Now it is starting to get exciting. It is time to get this Android going, but before you can do so, you need to install and configure a few tools, including SDKs:

✔ **Java JDK:** Lays the foundation for the Android SDK.

✔ **Android SDK:** Provides access to Android libraries and allows you to develop for Android.

✔ **Eclipse IDE (integrated development environment):** Brings together Java, the Android SDK, and the Android ADT (Android Development Tools), and provides tools for you to write your Android programs.

✔ **Android ADT:** Does a lot of the grunt work for you, such as creating the files and structure required for an Android app.

In the following sections, I show you how to acquire and install all these tools.

A benefit of working with open-source software is that most of the time, the tools you need to develop the software are free. Android is no exception to that rule. All the tools that you need to develop rich Android applications are free of charge.

The Android SDK changes *fast.* The installation instructions are accurate as of August 2011. If you run into any issues during the installation of the tools or frameworks, please consult the Android SDK Installation page as the installation may have changed since the writing of this book (`http://developer.android.com/sdk/installing.html`).

Getting the Java Development Kit

For some reason, the folks responsible for naming the Java SDK decided that it would be more appropriate to name it the Java Development Kit, or JDK for short.

Installing the JDK can be a somewhat daunting task, but I guide you through it one step at a time.

Downloading the JDK

Follow these steps to install the JDK on a Windows machine:

1. **Point your browser to** `www.oracle.com/technetwork/java/javase/downloads/index.html`.

    The Java SE Downloads page appears.

2. **Click the JDK link under the Java Platform (JDK) heading (see Figure 4-2).**

    If you're on a Mac, install the JDK through the Software Update panel.

A new Java SE Downloads page appears, asking you to specify which platform (Windows, Linux, or Mac) you'll be using for your development work.

Figure 4-2:
Select JDK.

The JDK link

3. **Using the Platform drop-down list, confirm your platform, and then click the Download button.**

   An optional Log in for Download screen appears.

4. **Click the Skip This Step link at the bottom of the page.**

5. **Click `JDK-6u24-windows-i586.exe` to download the file.**

   Windows opens a message box with a security warning. Note, this may be different if an updated version of Java has been released since the writing of this book.

6. **In the Save As dialog box, select the location where you want to save the file and click Save.**

The Web page shown in Figure 4-2 may look different in the future. To ensure that you're visiting the correct page, visit the Android SDK System Requirements page in the online Android documentation for a direct link to the Java SDK download page. View the requirements page at `http://developer.android.com/sdk/requirements.html`.

You must remember what version of the Java SDK you need to install. At the time of this writing, Android 3.0 supports Java SDK versions 5 and 6. If you install the wrong version of Java, you'll get unexpected results during development.

Installing the JDK

When the download is complete, double-click the downloaded file to install the JDK. You are prompted by a dialog box that asks whether you want to allow the program to make changes to your computer. Click the Yes button. (If you click the No button, the installation stops.) When you're prompted to do so, read and accept the license agreement.

That's all there is to it! You have the JDK installed and are ready to move to the next phase. In this section, I show you how to install the Android SDK step by step.

Getting the Android SDK

The Android SDK is composed of a debugger, Android libraries, a device emulator, documentation, sample code, and tutorials. You can't develop Android apps without the Android SDK.

To download the Android SDK, follow these steps:

1. **Point your browser to** `http://developer.android.com/sdk/index.html`.

2. **Choose the latest version of the SDK starter package for your platform. If you're using Windows, please select the recommended download as it has an installer to streamline the installation process for you.**

3. **Extract the SDK archive (.zip or .tgz).**

   I recommend extracting `/Library/android` on Mac OS X because I reference this location later in this chapter.

   You've just downloaded the Android SDK.

4. **Navigate to the directory where you extracted the SDK, and double-click the SDK Manager.exe file. (In Mac OS X, double click the android file in the tools directory.)**

5. **If you're prompted to accept the authenticity of the file, click Yes.**

   The Android SDK and AVD Manager dialog box opens.

6. **Select the SDK Platform Android 3.0 check box.**

For the purposes of this book, select version 3.0, as shown in Figure 4-3. At this writing, 3.0 is the latest and greatest version of Android. Check the boxes for the documentation and samples that correspond with Android version 3.0 (API 11).

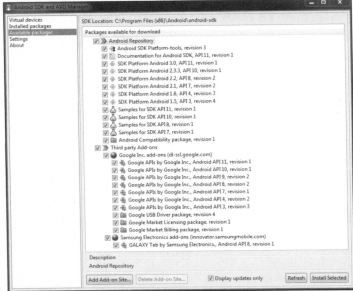

Figure 4-3: Choose packages to install.

Every time a new version of the Android OS is released, Google also releases an SDK that contains access to the added functionality in that version. If you want to include Bluetooth functionality in your app, for example, make sure that you have selected Android SDK version 2.0 or later, because this functionality isn't available in earlier versions.

7. **Click Install Selected.**

    The Choose Packages to Install dialog box opens. (See Figure 4-4.)

8. **Select the Accept radio button to accept the license and then click Install.**

9. **In the next dialog box, select Accept and click Install.**

    The Installing Archives dialog box opens, displaying a progress bar.

10. **When the archives installation is complete, click the Close button.**

While the Android SDK is attempting to connect to the servers to obtain the files, you may occasionally receive a `Failure to fetch URL` error. If this happens to you, navigate to Settings, select Force https://... Sources to be Fetched Using http://, and then attempt to download the available packages again.

Adding the Android NDK

The Android Native Development Kit (NDK) is a set of tools that allows you to embed components that use native code — code that you've written in a native language such as C or C++.

If you decide to take on the NDK, you still have to download the SDK. The NDK isn't a replacement for the SDK; it's an added functionality set that complements the SDK. You can learn more about the NDK here: `http://d.android.com/sdk/ndk/`.

Figure 4-4:
The Choose Packages to Install dialog box.

Getting the Total Eclipse

Now that you have the SDK, you need an integrated development environment (IDE) to use it. It's time to download Eclipse!

Choosing the right Eclipse version

Downloading the correct version of Eclipse is very important. At this writing, Android supports Eclipse versions Galileo and above. Check the Android System Requirements page at `http://developer.android.com/sdk/requirements.html`. If you're still unsure, download Eclipse Galileo (version 3.5), that's the version I'm currently using. When you download the file, you'll probably need to find the Older Versions link on the download page and select the latest Galileo version.

To download the correct version, navigate to the Eclipse downloads page (`www.eclipse.org/downloads`), select Eclipse IDE for Java Developers. Eclipse IDE for JAVA EE Developers works as well.

Installing Eclipse

Eclipse is a self-contained executable file; after you unzip it, the program is installed. Even though you could stop here, I suggest pinning a shortcut to your Start menu if you're on Windows so that Eclipse is easy to find when you need it.

To install Eclipse, you need to extract the contents of the Eclipse `.zip` file to the location of your choice. For this example, I use `C:\Program Files\Eclipse` when on Windows. For OS X users, unzip the .gz file and simply drag the application into your applications folder.

To install Eclipse on Windows, follow these steps:

1. **Double-click the shortcut that you just created to run Eclipse.**

   If you're running a recent version of Windows, the first time you run Eclipse, a Security Warning dialog box may appear. This dialog box tells you that the publisher has not been verified and asks whether you still want to run the software. Clear the Always Ask Before Opening This File check box, and click the Run button.

2. **Set your workspace.**

   When Eclipse starts, the first thing you see is the Workspace Launcher dialog box, as shown in Figure 4-5. Here, you can modify your workspace if you want, but for this book, I stick with the default:

   ```
   c:\users\<username>\workspace
   ```

   Leave the Use This as the Default and Do Not Ask Again check box deselected, and click the OK button.

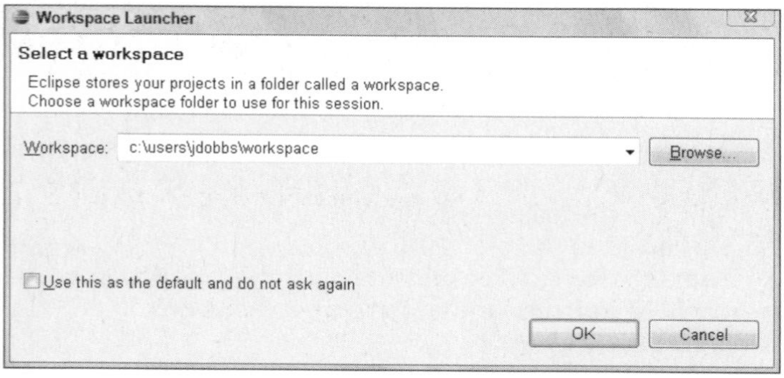

Figure 4-5:
Set your
workspace.

If you plan to develop multiple applications, I recommend using a sepa-rate workspace for each project. If you store multiple projects in one workspace, it gets difficult to keep things organized, and it's easy to change a similarly named file in a different project. Keeping projects in their own workspaces makes it easier to find the project when you have to go back to it to fix bugs.

When Eclipse finishes loading, you see the Eclipse welcome screen, shown in Figure 4-6.

Click to go to the workbench

Figure 4-6:
The Eclipse
welcome
screen.

3. **Click the curved-arrow icon on the right side of the screen to go to the workbench.**

Eclipse is installed and easily accessible. I show you how to add the Android Development Tools in the next section.

Configuring Eclipse

Android Development Tools (ADT) adds functionality to Eclipse to do a lot of the work for you. The ADT allows you to create new Android projects easily; it creates all the necessary base files so that you can start coding your application quickly. The ADT also allows you to debug your application by using the Android SDK tools. Finally, the ADT enables you to export a signed application file, known as an Android Package (APK), right from Eclipse, eliminating the need for some command-line tools. In the beginning, I had to use various command-line utilities to build an APK. Although that wasn't hard, it was tedious and sometimes frustrating. The ADT eliminates this frustrating process by guiding you through it, in "wizard format," from within Eclipse. I show you how to export a signed APK in Chapter 10.

Setting up Eclipse with the ADT

To set up Eclipse with the ADT (install ADT into Eclipse), follow these steps:

1. **Start Eclipse, if it's not already running.**

2. **Choose Help⇨Install New Software.**

   The Install window pops up (see Figure 4-7). This window lets you install new plug-ins in Eclipse.

3. **Click the Add button to add a new site**

   The Add Repository window appears (see Figure 4-8).

   Sites are the Web addresses where the software is hosted on the Internet. Adding a site to Eclipse makes it easier for you to update the software after a new version is released.

4. **Type a name in the Name field.**

   I recommend using Android ADT, but you can choose any name you like.

5. **Type** https://dl-ssl.google.com/android/eclipse/ **in the Location field.**

6. **Click the OK button.**

   Android ADT is selected in the Work With drop-down menu, and the available options are displayed in the Name and Version window of the Install Details dialog box.

7. **In the Install Details dialog box, select the check box next to Developer Tools and click the Next button.**

   The screen should look similar to Figure 4-9, but with different version numbers.

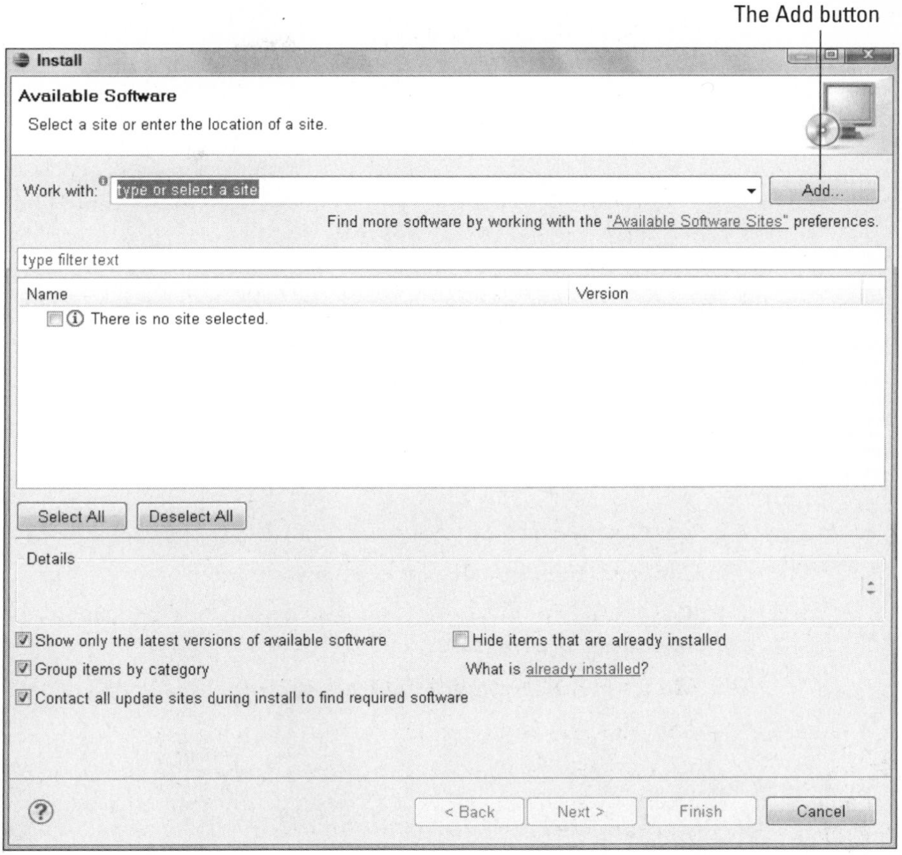

The Add button

Figure 4-7:
Click the
Add button
to add a
new site.

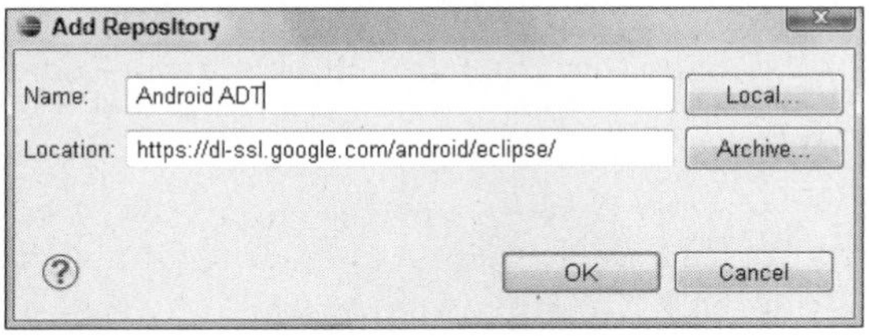

Figure 4-8:
Enter the
name and
location of
the site.

The Install Details dialog box should list both the Android Dalvik Debug Monitor Server (DDMS; see "Get physical with a real Android device," later in this chapter) and the ADT. Again, the version numbers will be different when you do this.

8. Click the Next button to review the software licenses.

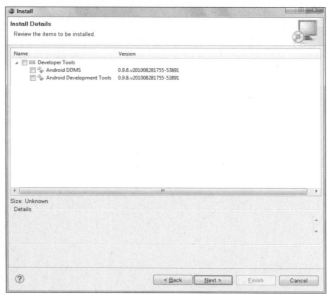

Figure 4-9:
Select
Developer
Tools.

9. **Click the Finish button.**

10. **When you're prompted to do so, click the Restart Now button to restart Eclipse.**

The ADT plug-in is installed.

Setting the location of the Android SDK

In this section, I guide you through the Android SDK configuration process. I know that this seems like a lot to do, but you're almost done, and you have to do this work only once. Follow these steps:

1. **Choose Window➪Preferences for Windows and Eclipse➪Preferences.**

The Preferences dialog box opens (see Figure 4-10).

2. **Select Android in the left pane.**

3. **For Windows, set the SDK Location to C:\android\android-sdk-windows. For OS X, set the SDK Location to /Library/android.**

4. **Click OK.**

Eclipse is configured, and you're ready to start developing Android apps.

If you're having difficulty downloading the tools from `https://dl-ssl.` `google.com/android/eclipse`, **try removing the** *s* **from** `https://`, as follows: `http://dl-ssl.google.com/android/eclipse`.

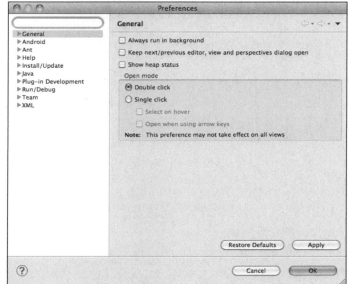

Figure 4-10:
Specify
the loca-
tion of the
SDK in the
Preferences
dialog box.

Getting Acquainted with the Android Development Tools

Now that the tools of the trade are installed, I introduce you to the SDK and some of the tools that are included with it.

Navigating the Android SDK

Whoa! You find a lot of folders in the SDK! Don't worry; the folder structure of the Android SDK is pretty easy to understand when you get the hang of it. You need to understand the structure of the SDK to master it. Table 4-1 outlines what each folder is and what it contains. Open the SDK with Windows Explorer or Mac Finder.

Table 4-1	Folders in the Android SDK
SDK Folder	*Description*
`extras`	Contains the drivers for Android devices. If you connect your Android device to the computer, you need to install this driver so that you can view, debug, and push applications to your phone via the ADT.
`tools`	Contains various tools that are available for use during development — debugging tools, view-management tools, and build tools, to name a few.
`temp`	Provides a temporary swap for the SDK. At times, the SDK may need a temporary space to perform some work. This folder is where that work takes place.
`samples`	Contains a bunch of sample projects for you to play with. Full source code is included.
`platforms`	Contains the platforms that you target when you build Android applications, such as folders named `android-8` (which is Android 2.2), `android-4` (which is Android 1.6), and so on.
`platform-tools`	Contains common tools specific to Android development.
`docs`	Contains a local copy of the Android SDK documentation.
`add-ons`	Contains additional APIs that provide extra functionality. This folder is where the Google APIs reside; these APIs include mapping functionality. This folder remains empty until you install any of the Google Maps APIs.

Targeting Android platforms

Android platform is just a fancy way of saying *Android version*. At this writing, seven versions of Android are available, ranging from version 1.1 through version 3.0. You can target (use framework version) any platform that you choose.

Figure 4-11 shows the percentage of each platform in use as of July 1, 2010. To view the current platform statistics, visit `http://developer.android.com/resources/dashboard/platform-versions.html`.

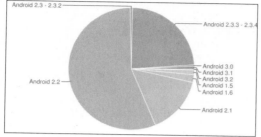

Using SDK tools for everyday development

You just installed the SDK tools. Now I introduce you to these tools so that they can work for you. The SDK tools are what you use to develop your Android apps. These tools let you develop applications easily as well as give you the ability to debug them. New features packed into every release enable you to develop for the latest version of Android.

Say hello to my little emulator

The emulator has to be my favorite tool of all. Not only does Google provide the tools you need to develop apps, but it also gives you this awesome little emulator that allows you to test your app! The emulator simulates testing environments, allowing you to test your app in a number of different screen sizes, resolutions, and platforms. The emulator comes in handy for testing your app at different screen sizes and resolutions. Having several devices connected to your computer at the same time is not always possible, but you can run multiple emulators with varying screen sizes and resolutions.

The emulator does have some limitations, however — it cannot emulate certain hardware components such as the accelerometer. When you're developing an app that uses Bluetooth, for example, use an actual device that has Bluetooth. If you develop on a speedy computer, testing on an emulator is fast, but on slower machines, the emulator can take a long time to do a seemingly simple task. When I develop on an older machine, I usually use an actual device, but when I use my newer, faster machine, I typically use the emulator because I don't notice much lag, if any. Still, these limitations are minor. Plenty of apps can be developed and tested using only an emulator. Some limitations include testing wireless connectivity, Bluetooth interaction, and various other hardware dependent features.

Get physical with a real Android device

The emulator is awesome, but sometimes you need to test on an actual device. The DDMS allows you to debug your app on an actual device, which comes in handy for developing apps that use hardware features that aren't or can't be emulated. (See the section, "Debug your work," later in this chapter.) Suppose that you're developing an app that tracks the user's location. You can send coordinates to the device manually, but at some point in your development, you probably want to test the app and find out whether it in fact displays the correct location. Using an actual device is the only way to do this.

If you develop on a Windows machine and want to test your app on a real device, you need to install a USB driver. (If you're on a Mac or Linux machine, you can skip this section, because you don't need to install the USB driver.) To download the Windows USB driver for Android devices, follow these steps:

1. **In Eclipse, choose Window➪Android SDK and AVD Manager.**

   The Android SDK and AVD Manager dialog box opens.

2. **In the left pane, select Available Packages. (See Figure 4-12.)**

3. **Expand the Android repository and select the Third Party Add-ons package.**

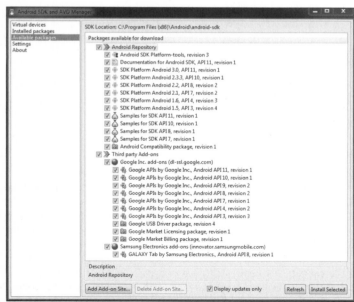

Figure 4-12:
The available packages.

4. **Click the Install Selected button.**

   The Choose Packages to Install dialog box opens. (See Figure 4-13.)

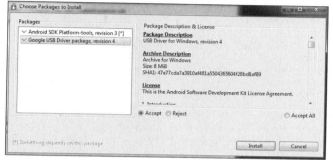

5. **Select the Accept radio button to accept the license and then click the Install button.**

   The Installing Archives dialog box opens, displaying a progress bar.

6. **When the package finishes downloading and installing, click the Close button.**

7. **Exit the Android SDK and AVD Manager dialog box.**

Debug your work

The DDMS equips you with the tools you need to find those pesky bugs. You can go behind the scenes as your app is running to see the state of hardware such as wireless radios. But wait. There's more! It also simulates actions that you normally need an actual device to do, such as sending Global Positioning System (GPS) coordinates manually, simulating a phone call, or simulating a text message. Get all the DDMS details at `http://developer.android.com/guide/developing/tools/ddms.html`.

Try out the API and SDK samples

The API and SDK samples are provided to demonstrate how to use the functionality provided by the API and SDK. If you ever get stuck and can't figure out how to make something work, visit `http://developer.android.com/resources/samples/index.html`. Here, you can find code samples of almost anything from using Bluetooth to making a two-way text application or a 2D game.

You also have a few samples in your Android SDK. Simply open the Android SDK and navigate to the `samples` directory, which contains various samples that range from interacting with services to manipulating local databases. Spend some time playing with the samples. The best way to learn Android is to look at existing working code bases and then play with them in Eclipse.

Give the API demos a spin

The API demos inside the `samples` folder in the SDK are a collection of apps that demonstrate how to use the included APIs. Here, you can find sample apps with a ton of examples, such as these:

- ✔ Notifications
- ✔ Alarms
- ✔ Intents
- ✔ Menus
- ✔ Search
- ✔ Preferences
- ✔ Background services

If you get stuck or just want to prep yourself for writing your next spectacular Android application, check out the complete details at `http://developer.android.com/resources/samples/ApiDemos/index.html`.

Chapter 5

Your First Android Project

You're excited to get started building the next best Android application known to man, right? Good! But hold on — don't get ahead of yourself. In this chapter, I walk you through the process of creating your first Android application, a very simple Hello Android application. This app demonstrates a few key aspects of the development process and requires no coding at all. (What? No coding? How's that possible? Follow along as I show you.) Building it will get you up and running as an app developer. Just go one step at a time — with a little practice, you'll be creating that blockbuster Android app in no time.

Starting a New Project in Eclipse

First things first: You need to start Eclipse. (Refer to Chapter 4.) After you do, your screen should look similar to the one in Figure 5-1. Now you're ready to start cooking with Android.

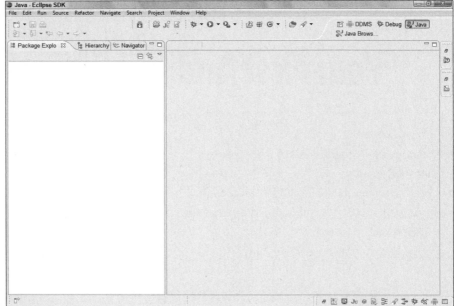

Figure 5-1:
The Eclipse
develop-
ment
environ-
ment.

In Chapter 4, I show you how to install the Eclipse Android Development Tools (ADT) plug-in. The ADT plug-in gives you the power to generate new Android applications directly from within the Eclipse File menu. That's exactly what you're about to do. To create your first Android Application project, follow these steps:

1. **In Eclipse, choose File⇨New⇨Project.**

   The New Project/Select a Wizard dialog box opens, as shown in Figure 5-2.

2. **From the New Project/Select a Wizard dialog box, expand the Android item by clicking the Android folder.**

3. **Click Android Project within the expanded folder and then click the Next button.**

   The New Android Project dialog box appears, as shown in Figure 5-3.

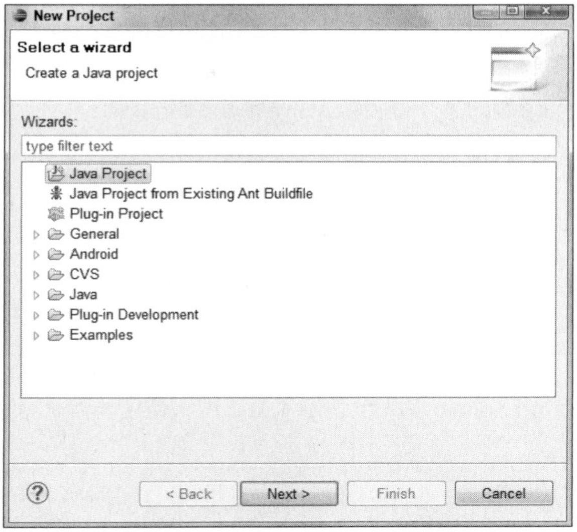

Figure 5-2:
The New
Project/
Select a
Wizard
dialog box.

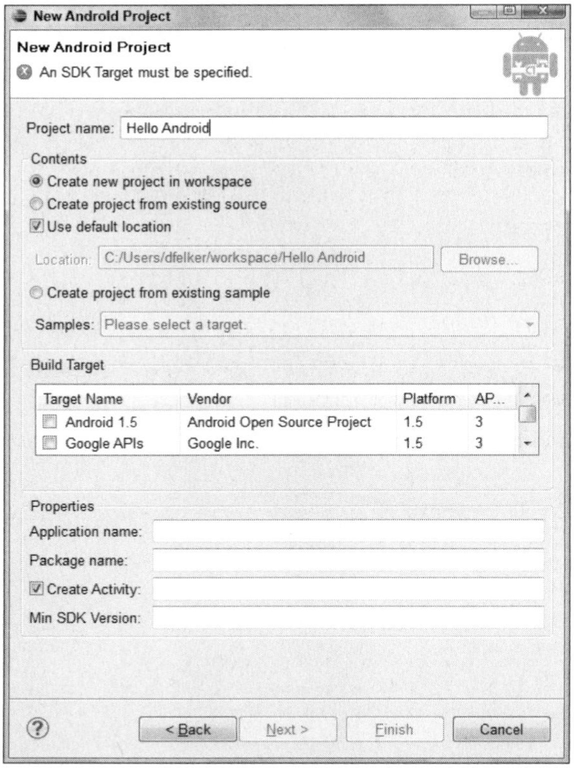

Figure 5-3:
The New
Android
Project
dialog box.

4. **In the Project Name field, type** Hello Android.

   The Project Name field is very important because the descriptive name you provide identifies your project in the Eclipse workspace. After your project is created, a folder in the workspace is named with the project name you define here.

5. **In the Contents panel, leave the default radio button, Create New Project in Workspace, and the default check box, Use Default Location, selected.**

   These defaults are selected automatically when a new project is created. The Contents panel identifies where the contents of your Eclipse projects are going to be stored in the file system. The contents are the source files that make up your Android project.

   When you set up Eclipse in Chapter 4, the Eclipse system asked you to set your default workspace, usually your home directory (c:\users\<username>\workspace). A *home directory* is where Eclipse places files pertinent to your project. Figure 5-4 shows my home directory.

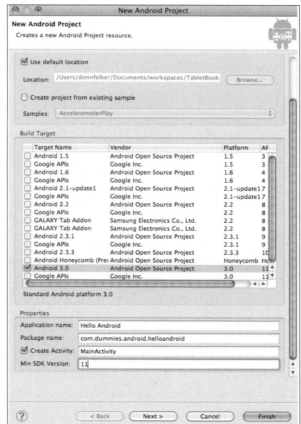

Figure 5-4:
My default workspace location for the Hello Android project is C:/Users/dfelker/workspace.

TIP

If you would rather store your files in a location other than the default workspace location, deselect the Use Default Location check box. Doing this enables the Location text box. Click the Browse button and select a location where you want your files stored.

6. **In the Build Target section, select Android 3.0.**

The Build Target section identifies which application programming interface (API) you want to use to develop this project. Selecting the Android 3.0 framework enables you to develop with the Android 3.0 APIs, which include new features such as the holographic UI, Fragments, and Action Bar APIs. If you selected Android 1.6 as the target, you'd have access only to version 1.6 APIs, and you would not be able to use any features supported by version 3.0 (or 2.2, or 2.3, and so on). Only the features in the targeted framework are supported. If you installed other software development kits (SDKs) in Chapter 4, you might have the option of selecting them at this point.

For more information, see the section "Understanding the Build Target and Min SDK Version settings," later in this chapter.

7. **In the Properties section, type** Hello Android **in the Application Name box.**

The application name is the name of the application as it pertains to Android. When the application is installed on the emulator or physical device, this name appears in the application launcher.

Java package nomenclature

A *package* in Java is kind of like a module: It's Java's way of organizing Java classes into namespaces. Each package must have a unique name for the classes it contains. Classes in the same package can access one another's package-access members.

Packages enable you to keep your code organized. You could use a Java package, for example, for all your Web-related communications. Any time you needed to find one of your Web-related Java classes, you could open that Java package and work on your Web-related Java classes.

Java packages have a naming convention defined as the hierarchical naming pattern.

Each level of the hierarchy is separated by periods. A package name starts with the highest-level domain name of the organization; then the subdomains are listed in reverse order. At the end of the package name, the company can choose what it would like to call the package. The package name `com.dummies.android.helloandroid` is the name you will use for this example.

Notice that the highest-level domain is at the front of the package name (`com`). Subsequent subdomains are separated by periods. The package name traverses down through the subdomains to get to the final package name of `helloandroid`.

8. **In the Package Name box, type** com.dummies.android.helloandroid.

   This is the name of the Java package (see the nearby sidebar "Java package nomenclature").

9. **Check the Create Activity check box (if it isn't already checked). In the Create Activity text box, type** MainActivity.

   The Create Activity section defines what the initial activity will be called. This step is the entry point to your application. When Android runs your application, this file is accessed first. A common naming pattern for the first activity in your application is `MainActivity.java` (how creative, right?).

10. **In the Min SDK Version box, type** 11.

    The Min SDK Version defines the minimum version code of Android that the user must have before he can run your application. Note that this field is not required to create your app.

    For more information, see the section "Understanding the Build Target and Min SDK Version settings," later in this chapter.

11. **Click the Finish button.**

    You're done! You should see Eclipse with a single project in the Package Explorer, as shown in Figure 5-5.

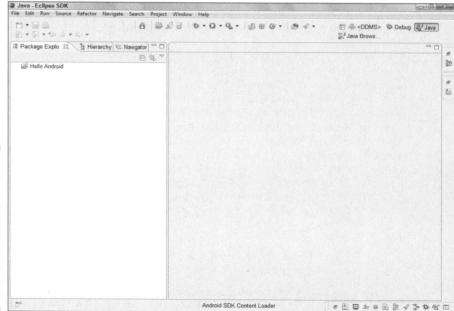

Figure 5-5:
The Eclipse development environment with your first Android project, Hello Android.

Understanding Android versioning

Version codes are not the same as version names. Huh? Android has both version names and version codes. Each version name has one and only one version code associated with it. The following table outlines the version names and their respective version codes.

You can also find this information in the Build Target section of the New Android Project dialog box.

Version Name (Platform Level)	Version Code (API Level)
1.5	3
1.6	4
2.0	5
...	...
3.1	12

Deconstructing Your Project

The Hello Android project generated by Eclipse is a fresh, clean project with no compiled binary sources. You also need to understand what happens under the hood of Eclipse at a high level. I cover this information in the next few sections.

Responding to error messages

If you were quick enough to look (or if your computer runs on the slower edge of the spectrum), you may have noticed a little red icon hovering over the Hello Android folder icon in the Package Explorer in your Eclipse window after you clicked the Finish button. If you didn't see it, you can see an example in Figure 5-6. Displaying that icon is how Eclipse lets you know that something is wrong with the project in the workspace.

By default, Eclipse is set up to let you know when an error is found within a project with this visual queue. Behind the scenes, Eclipse and the Android Development Tools are doing a few things for you:

✓ **Providing workspace feedback:** This feedback lets you know when a problem exists with any of the projects in the workspace. You receive notification in Eclipse through icon overlays, such as the one shown in Figure 5-6. Another common icon overlay is the small yellow warning icon, which alerts you to some warnings in the contents of the project.

✓ **Automatically compiling:** By default, Eclipse autocompiles the applications in your workspace when any files within them are saved after a change.

TIP

If you don't want automatic recompilation turned on, choose Project⇨ Build Automatically to disable the automatic building of the project. If this option is deselected, you need to build your project manually by pressing Ctrl+B each time you change your source code.

By now you're probably wondering: How can I have an error with this project? I just created it through the New Android Project wizard, so what gives? When the project was added to your workspace, Eclipse took over, and in conjunction with the ADT, it determined that the project in the workspace had an error: The gen folder and all its contents were missing. (I cover the gen folder in "Understanding Project Structure," later in this chapter.)

The gen folder is automatically generated by Eclipse and the ADT when the compilation takes place. As soon as the New Android Project wizard was completed, a new project was created and saved in Eclipse's workspace. Eclipse recognized this fact and said, "Hey! I see some new files in my workspace. I need to report any errors I find as well as compile the project." Eclipse reported the errors by placing an error icon over the folder. Immediately thereafter, the compilation step took place. During the compilation step, the gen folder was created by Eclipse, and the project was successfully built. Then Eclipse recognized that the project did not have any more errors. At that time, it removed the error icon from the folder, leaving you with a clean workspace and a clean folder icon, as shown in Figure 5-5.

A project with errors

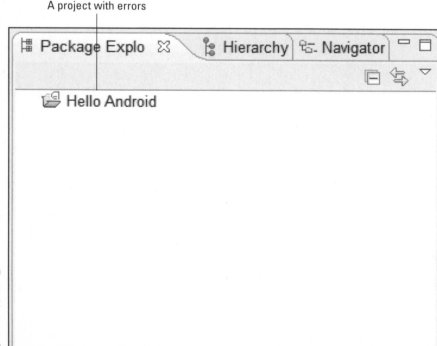

Figure 5-6:
A project
with errors
in Eclipse.

Understanding the Build Target and Min SDK Version settings

So how does the Build Target setting differ from the Min SDK Version setting?

The *build target* is the operating system you use to write code. If you choose 3.0, you can write code with all the APIs in version 3.0. If you choose 1.6, you can write code only with the APIs that are in version 1.6. You can't use the Action Bar APIs in version 2.2, for example, because they weren't introduced until version 3.0. If you're targeting 3.2, though, you can write with the Action Bar APIs.

You should know which version you want to target before you start writing your Android application. Identify which Android features you need to use to ensure that your app will function as you expect. If you're positive that you're going to need Bluetooth support, you need to target at least version 2.0. If you're not sure which versions support the features you're looking for, you can find that information on the platform-specific pages in the SDK section of `http://d.android.com`. The Android 3.0 platform page is at `http://d.android.com/sdk/android-3.0.html`.

Android operating-system (OS) versions are backward-compatible. If you target Android version 1.6, for example, your application can run on Android 3.0, 2.3.3, 2.2, . . . , and of course 1.6. The benefit of targeting the 1.6 framework is that your application is exposed to a much larger market share. Your app can be installed on 1.6, 2.0, 2.1, and so on, all the way up to 3.0 devices (and future versions, assuming that no breaking framework changes are introduced in future Android OS releases). Selecting an older version doesn't come without consequences, however. By targeting an older framework, you're limiting the functionality that you can access. By targeting 1.6, for example, you won't have access to the Bluetooth APIs.

The Min SDK Version setting is the minimum version of Android that the user must be running for the application to run properly on his or her device. This field isn't required to build an app, but I highly recommend that you fill it in. If you don't indicate the Min SDK Version, a default value of 1 is used, indicating that your application is compatible with all versions of Android.

If your application is *not* compatible with all versions of Android (such as if it uses APIs that were introduced in version code 5 — Android 2.0), and you haven't declared the Min SDK Version, when your app is installed on a system with an SDK version code of less than 5, your application will crash at run time when it attempts to access the unavailable APIs. As best practice, always set the Min SDK Version in your application to prevent these types of crashes.

Version codes and compatibility

The Min SDK Version is also used by the Android Market (which I cover in detail in Chapter 10) to help identify which applications to show you based on which version of Android you're running. If your device is running version code 3 (Android 1.6), you would want to see the apps pertinent to your version, not version code 11 (Android 3.0) apps. The Android Market manages which apps to show to each user through the Min SDK Version setting.

If you're having trouble deciding which version to target, the current version distribution chart can help you decide. That chart is located here: `http://developer.android.com/resources/dashboard/platform-versions.html`.

A good rule is to analyze the distribution chart at `http://developer.android.com` to determine which version will give your app the best market share. The more devices you can target, the wider the audience you will have, and the more installs you have, the better your app is doing.

Setting Up an Emulator

Aw, shucks! I bet you thought you were about to fire up your new app. Well, you're almost there. You have one final thing to cover, and then you get to see all of your setup work come to life in your Hello Android application. To see this application in a running state, you need to know how to set up an emulator through the various different launch configurations.

First, you need to create an Android Virtual Device (AVD), also known as an emulator. An AVD looks, acts, walks, and talks (well, maybe not walks and talks) just like a real Android device. AVDs can be configured to run any particular version of Android as long as the SDK for that version is downloaded and installed.

It's time to get reacquainted with your old buddy the Android SDK and AVD Manager. Follow these steps to create your first AVD:

1. **To open the Android SDK and AVD Manager, click the icon on the Eclipse toolbar shown in Figure 5-7.**

The SDK/AVD Manager

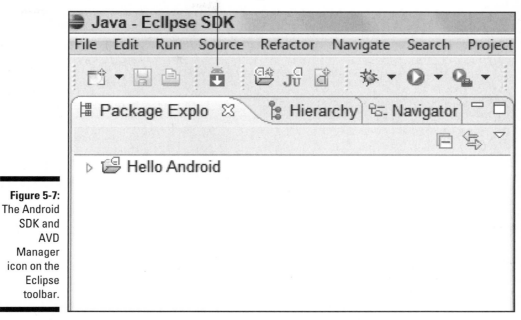

Figure 5-7:
The Android
SDK and
AVD
Manager
icon on the
Eclipse
toolbar.

The Android SDK and AVD Manager opens. (See Figure 5-8.)

Figure 5-8:
The Android
SDK and
AVD
Manager
dialog box.

2. **Click the New button.**

The Create New Android Virtual Device (AVD) dialog box opens, as shown in Figure 5-9.

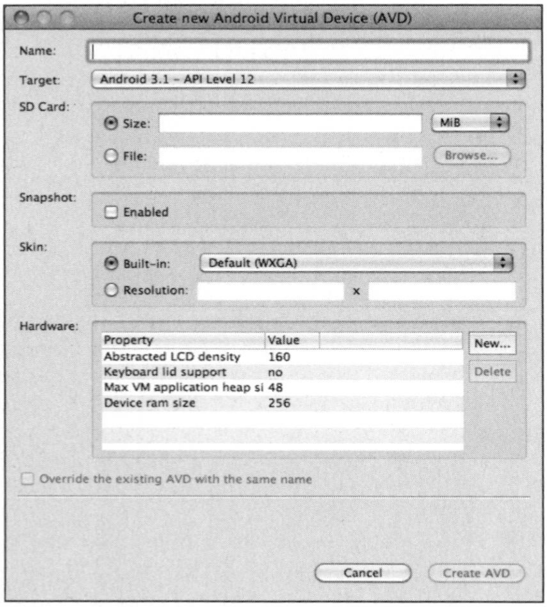

Figure 5-9:
The Create
New
Android
Virtual
Device
(AVD) dialog
box.

3. **For this AVD, in the Name field, type** 3_0_Default_WXGA.

Be careful when naming your AVDs. Android is available on many devices in the real world, such as phones, e-book readers, and netbooks. A time will come when you have to test your app on various configurations; therefore, adhering to a common nomenclature when creating your AVDs can later help you recognize which AVD is for what purpose. The nomenclature I tend to follow is the following:

```
{TARGET_VERSION}_{SKIN}_{SCREENSIZE}[{_Options}]
```

In the previous step, I suggest you use the name 3_0_Default_WXGA. If you do this, the AVD will have a TARGET_VERSION of Android 3.0. (The version name 3.0 is transformed into 3_0. The underscores are used in place of periods to keep the name of the AVD combined. Creating an AVD name as a single combined word helps when you're working in advanced scenarios with AVDs via the command line.)

The SKIN is the name of the skin of the emulator. Emulators can have various skins that make them look like actual devices. The default skin (Default) is provided by the Android SDK.

The SCREENSIZE value is the size of the screen with regard to the Video Graphics Array (VGA) size. The default is WXGA. Other options include QVGA and WVVGA800.

4. **In the Target box, select Android 3.0 — API Level 11.**

5. **In the SD Card section, leave the fields blank.**

You have no use for an SD Card in this application. You would use the SD Card option if you needed to save data to the SD Card. If you want to have an SD Card emulator in the future, insert the size of the SD Card in megabytes (MB) that you would like to have created for you. At that time, an emulated SD Card will be created and dropped in your local file system.

6. **Leave the Skin option set to Built-in (WXGA).**

7. **Change the Device RAM size in the Hardware section.**

The Hardware section outlines the hardware features your AVD should emulate. The Android 3.0 emulator you're building emulates an ARM Processor, and to emulate this processor on your machine takes a lot of RAM. Although the default value of 256 should suffice, if you bump this up to a larger value the emulator will run much faster. If your machine has more than 2 GB of RAM, bump this value up to 1024 (1 GB of RAM) for the emulator. Doing this makes the emulator a little more snappy and responsive.

8. **Click the Create AVD button.**

The Android SDK and AVD Manager dialog box should now display the new AVD. (See Figure 5-10.)

Figure 5-10: The recently created AVD in the Android SDK and AVD Manager dialog box.

9. **Close the Android SDK and AVD Manager dialog box.**

You've created your first Android virtual device. Congratulations!

If you receive an error message after you create your AVD that says `Android requires .class compatibility set to 5.0. Please fix project properties.`, you can fix it by right-clicking the project in Eclipse and choosing Android Tools⇨Fix Project Properties from the context menu.

Creating Launch Configurations

You're almost at the point where you can run the application. A launch configuration specifies the project to run, the activity to start, and the emulator or device to connect to. Whoa! That's a lot of stuff happening real quickly. Not to worry; the ADT can help you by automating a lot of the key steps so that you can get up and running quickly.

The Android ADT gives you two options for creating launch configurations:

- ✔ **Run configuration:** Used when you need to run your application on a given device. You use run configurations most of the time during your Android development career.
- ✔ **Debug configuration:** Used for debugging your application while it's running on a given device. (You won't need to use this now. Debugging is discussed in Chapter 7.)

When you first run a project as an Android application by choosing Run⇨Run, the ADT automatically creates a run configuration for you. The Android Application option is visible when you choose Run⇨Run. After the run configuration is created, it's the default run configuration, used each time you choose Run⇨Run from then on.

Creating a run configuration

Now it's your turn to create a run configuration for your application.

If you're feeling ambitious and decide that you'd like to create a run configuration manually, follow along here. Don't worry — it's very simple. Follow these steps:

1. **Choose Run⇨Run Configurations.**

The Run Configurations dialog box opens, as shown in Figure 5-11. In this dialog box, you can create many types of run configurations. The left side of the dialog box lists many types of configurations, but the ones that apply to Android apps are as follows:

- *Android Application: Used to run an Android Application*
- *Android JUnit Test: Used to run a jUnit Unit Test against an Android Application*

2. **Select the Android Application item and click the New Launch Configuration icon, shown in Figure 5-11 (or right-click Android Application and choose New from the context menu).**

   The New Launch Configuration window opens.

3. **Type** ExampleConfiguration **in the Name field.**

4. **On the Android tab, select the project you are creating this launch configuration for. To do this, first click the Browse button.**

   The Project Selection dialog box opens.

5. **Select Hello Android and click the OK button (see Figure 5-12).**

New Launch Configuration icon

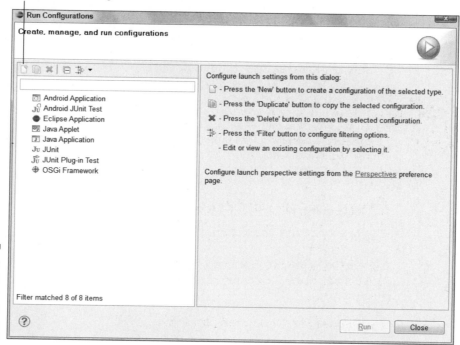

Figure 5-11:
The Run Configurations dialog box.

The Run Configurations dialog box reopens.

Figure 5-12:
Selecting
the project
for the new
launch con-
figuration.

6. **On the Android tab, leave the Launch Action option set to Launch Default Activity.**

   In this case, the default activity is `MainActivity`, which you set up in the section "Starting a New Project in Eclipse," earlier in this chapter.

7. **On the Target tab (see Figure 5-13), leave Automatic selected.**

8. **In the Select a Preferred Android Virtual Device for Deployment section, select the 3_0_Default_WXGA device.**

   This device is the AVD that you created previously. By selecting it, you're instructing this launch configuration to launch this AVD when a user runs the app by choosing Run⇨Run. This view has both manual and automatic options. The manual option allows you to choose which device to connect to when using this launch configuration. Automatic sets a predefined AVD to use when launching in this current launch configuration.

9. **Leave the rest of the settings alone, and click the Apply button.**

Congratulations! You've created your first launch configuration by hand.

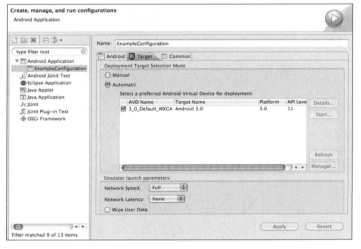

Figure 5-13:
A new,
manually
created
launch
configura-
tion named
Example-
Configura-
tion.

Duplicating your launch configuration for quick setup

At some point, during your very successful and lucrative Android develop-ment career, one of your applications may have a problem on one particular device. Launch configurations are designed to help you launch into a par-ticular environment quickly. Setting up launch configurations can be a time-consuming task, however — which can be frustrating, especially if the new launch configuration is only slightly different from an existing launch configu-ration. Fortunately, the ADT has included functionality that duplicates exist-ing launch configurations. This functionality enables you to quickly create various launch configurations that are set up independently with their own configuration in mind.

To duplicate an existing launch configuration, follow these steps:

1. **Choose Run⇨Run Configurations to open the launch configuration window.**

2. **In the left panel, right-click the configuration you want to duplicate (which, for this example, is `ExampleConfiguration`), and choose Duplicate from the context menu.**

   Doing this creates a new launch configuration that's an exact copy of `ExampleConfiguration`. Its name will be `Example-Configuration (1)`.

3. **Give the new run configuration a unique name. (For this example, type** DuplicateTest **in the Name field near the top of the window.)**

    At this point you can change various settings to give the launch configuration a unique configuration.

4. **(Optional) For the purposes of this chapter, you won't need the** DuplicateTest **launch configuration; it was created only to illustrate how to duplicate an existing launch configuration. If you'd like to delete this configuration, select** DuplicateTest **in the left panel and click the Delete button on the toolbar, or right-click it and choose Delete from the context menu.**

5. **Click the Close button to close the Run Configurations dialog box.**

Running the Hello Android App

Congratulations! You've made it! Understanding the basics of how to get an Android application up and running is a simple yet detailed process. You're now ready to see your hard work in action. You've created a launch configuration and an Android Virtual Device; now it's time for you to get the application running. Finally!

Running the application is simple. Just choose Run⇨Run or press Ctrl+F11 (but don't do that now!). Doing so causes the ADT to compile your application and launch it in an emulator using the default launch configuration you built earlier in this chapter — in this case, ExampleConfiguration.

However, the Android 3.0 WXGA emulator is *huge,* and unless you start the emulator first, running your app will bring up the emulator full size — a size well beyond the bounds of your screen, making the emulator impossible to use. This is because the screen density of the emulator matches that of your display. Because tablets are high density devices, the overall screen size will be very large, expanding beyond the length of your screen. When you start the emulator before running your app, however, you can define the size of its display to fit your screen. To start the emulator and launch your Hello Android app, follow these steps:

1. **Open the Android SDK Manager from the Eclipse menu on the toolbar.**

    This is the same step that you performed when you created the AVD in the "Setting Up an Emulator" section, earlier in this chapter.

2. **In the Android SDK Manager, select the 3_0_Default_WXGA AVD and then press the Start button.**

    Doing this opens the Launch Options screen, as shown in Figure 5-14.

Figure 5-14:
The Launch
Options
dialog box.

3. **Select the Scale Display to Real Size check box and place the value of 8 into the Screen Size (in) field. Notice the Scale label below the input fields now says 0.38, meaning that you have scaled the emulator size down to 38 percent of its original size. Press Launch and the emulator will now launch.**

As noted previously, the Android 3.0 emulator is a beast of an emulator. Because the emulator is emulating an ARM processor, it can take awhile for the emulator to boot up (especially if you have the minimum required RAM). I advise that you start the emulator when you open Eclipse so that by the time you're ready to test your application on the emulator, the emulator will be running and fully functional.

4. **If you didn't create a launch configuration, you see the Run As dialog box, shown in Figure 5-15. Choose Android Application and click OK, and a launch configuration is created for you. On the other hand, if you created `ExampleConfiguration` by following the steps in "Creating a run configuration," earlier in this chapter, you see the first emulator boot screen, as shown in Figure 5-16.**

Figure 5-15:
The Run As
dialog box.

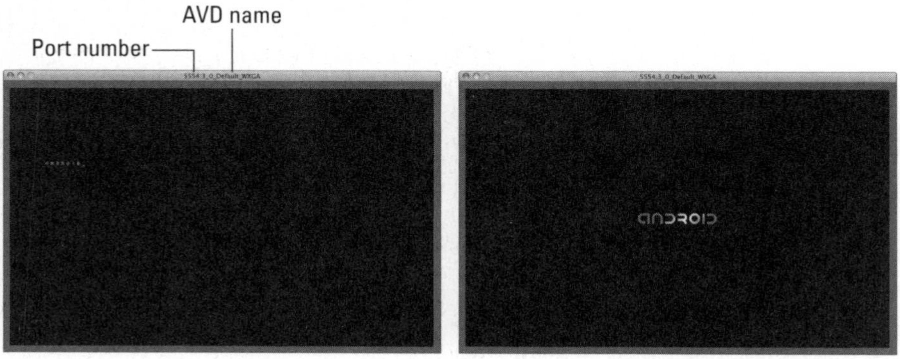

Port number ⎯ AVD name

Figure 5-16:
The first
and second
emulator
boot
screens.

Help! My emulator never loads! It stays stuck on one of the loading screens! No need to worry, comrade. The first time the emulator starts, the system could take upwards of 10 minutes for the emulator to finally finish loading. The reason for the lengthy loading time is that you're running a virtual Linux system in the emulator. The emulator has to boot up and initialize. The slower your computer, the slower the emulator will be in its boot process.

The emulator has two boot screens. The first contains the port number that the emulator is running on your computer (5554) and the AVD name (3_0_Default_WXGA). Roughly one half of boot time is spent on this screen. The port number is useful when you have multiple instances of the emulator running. You can use this port number as a reference to each emulator when using command-line tools. More information about this can be found here: http://d.android.com/guide/developing/tools/adb.html.

The second boot screen shows the Android logo. (Refer to Figure 5-16). This logo is the same one that default Android OS users see when they boot their phones (that is, unless a device manufacturer has installed its own user-interface customizations, as on the HTC Sense).

When the emulator completes its loading phase, the default home screen appears. (See Figure 5-17.)

Save valuable time by leaving the emulator running. The emulator doesn't have to be loaded each time you want to run your application. After the emulator is running, you can change your source code and then rerun your application. The ADT will find the running emulator and deploy your application to the emulator.

Figure 5-17:
The loaded
3_0_
Default_
WXGA
emulator.

Immediately after the emulator's default home screen appears, the ADT starts
the Hello Android application for you. You should see a black screen containing the words `Hello World, MainActivity!`, as shown in Figure 5-18.
Congratulations! You just created and started your first Android application.

Figure 5-18:
The Hello
Android
application
in the
emulator.

If your application doesn't start within five to ten seconds after the emulator's default home screen appears, simply run the application from Eclipse again by choosing Run➪Run. The application is redeployed to the device, and it starts running. Sometimes the ADT will time out when attempting to install your application onto the emulator. Timeouts often occur when the emulator is booting. If you're not sure what's going wrong, you can view the status of the installation via the Console view in Eclipse, as shown in Figure 5-19.

The Console

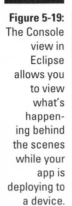

Figure 5-19:
The Console
view in
Eclipse
allows you
to view
what's
happen-
ing behind
the scenes
while your
app is
deploying to
a device.

Here's the full text of that information:

```
[2010-07-05 13:13:46 - Hello Android] ----------------------------
[2010-07-05 13:13:46 - Hello Android] Android Launch!
[2010-07-05 13:13:46 - Hello Android] adb is running normally.
[2010-07-05 13:13:46 - Hello Android] Performing com.dummies.android.
        helloandroid.MainActivity activity launch
[2010-07-05 13:13:46 - Hello Android] Automatic Target Mode: using existing
        emulator 'emulator-5554' running compatible AVD '3_0_Default_WXGA'
[2010-07-05 13:13:48 - Hello Android] Application already deployed. No need to
        reinstall.
[2010-07-05 13:13:48 - Hello Android] Starting activity com.dummies.android.
        helloandroid.MainActivity on device
[2010-07-05 13:13:49 - Hello Android] ActivityManager: Starting: Intent {
        act=android.intent.action.MAIN cat=[android.intent.category.
        LAUNCHER] cmp=com.dummies.android.helloandroid/.MainActivity }
[2010-07-05 13:13:49 - Hello Android] ActivityManager: Warning: Activity not
        started, its current task has been brought to the front
```

The Console view provides valuable information on the state of the application deployment. The content above demonstrates what happens when you try to deploy an application that has already been deployed but no changes have occurred since the last deployment. It lets you know it's launching an activity; shows what device the ADT is targeting; and shows warning information, as in the last line of the output. In that line, ADT informs you that the activity it attempted to launch in the previous line — `MainActivity`, in this case — hasn't been started because it was already running. Because the activity was already running, ADT brought that task to the foreground (the Android screen) for you to see.

Understanding Project Structure

Congratulations again! You created your first application. You even did it without coding. It's nice that the ADT provides you with the tools to fire up a quick application, but that's not going to help you create your next blockbuster application. The beginning of this chapter walks you through how to create a boilerplate Android application with the New Android Project wizard. From here on, you will use that application's file structure that the Android wizard created for you.

The following sections discuss the files and folders in this structure. You shouldn't skim these sections — trust me, they're important! — because you spend your entire Android development career navigating them. Understanding what they do and how they got there is a key aspect of understanding Android development.

Navigating the app's folders

In Eclipse, click the Package Explorer tab and this will display the Hello Android project's folder structure so that it resembles Figure 5-20.

When the Hello Android project's file structure is expanded, the list of subfolders includes

- ✔ **src:** This folder contains the Java source code for your project.
- ✔ **gen:** This folder contains generated code that Eclipse contains.
- ✔ **The Target Android library folder (in this case, `Android 3.0`):** This folder includes the `android.jar` file, which is the current version of Android.
- ✔ **assets:** This folder stores data files and other application "assets."
- ✔ **res:** This folder contains the bitmap, layout, and additional string resources your application utilizes.

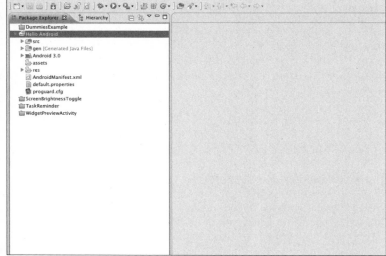

Figure 5-20:
The
Package
Explorer
with the
Hello
Android
project
folder
structure
expanded.

These folders aren't the only ones that you can have inside an Android project, but they're the default folders created by the New Android Project wizard. Other folders include bin, libs, and referenced libraries.

You won't see the bin folder initially, because in the latest version of the ADT it's hidden from view. (This may change in future versions of the ADT.) The libs and referenced libraries folders don't show up until you add a third-party library and reference it in your project. I cover this process in detail later in this chapter.

Also in this project are two files:

- **AndroidManifest.xml:** This file helps you identify the components that build and run the application.

- **default.properties:** This file helps you identify the default properties of the Android project (such as Android version).

I discuss all these folders and files in the following sections.

Source (src) folder

The source folder — known as the src folder in Android projects — includes your stub MainActivity.java file, which you created in the New Android Project wizard earlier in this chapter. To inspect the contents of the src folder, you must expand it. Follow these steps:

1. **Select the** `src` **folder, and click the small arrow to the left of the folder to expand it.**

   You see your project's default package: `com.dummies.android.` `helloandroid.`

2. **Select the default package, and expand it.**

   This step exposes the `MainActivity.java` file within the `com.dummies.` `android.helloandroid` package, as shown in Figure 5-21.

Figure 5-21:
The
expanded
src folder.

You aren't limited to a single package in your Android applications. In fact, separating different pieces of core functionality in your Java classes into packages is considered to be a best practice. For example, if you had classes whose responsibility was to communicate with a Web API through eXtensible Markup Language (XML), you could combine them into a single package. And if your application had several customer domain model classes represented by `Customer` objects that were retrieved through the Web API classes, you could combine those classes into a single package, too. Then you would have two additional Java packages in your Android app:

✔ **com.dummies.android.helloandroid.http:** Containing the HTTP-related Java classes (Web APIs)

✔ **com.dummies.android.helloandroid.models:** Containing the domain model Java classes

An Android project set up this way would look similar to Figure 5-22.

Target Android Library folder

I know — I skipped the `gen` folder! I delve into that folder later, when I discuss the `res` folder. For now, I want to focus on the Target Android Library folder, which isn't really a folder per se, but is more along the lines of an item in Eclipse generated as a symbolic link through the ADT. This item simply gives you a visual queue to what Android framework you're targeting.

New packages

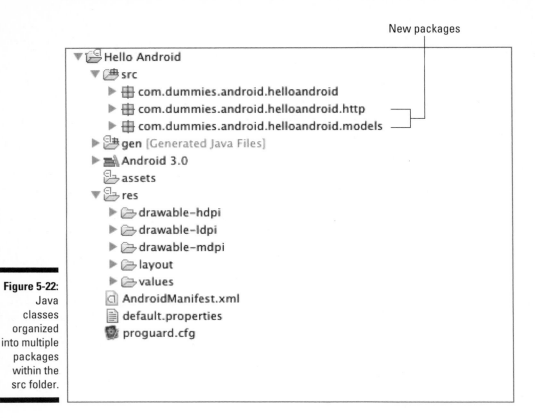

Figure 5-22:
Java
classes
organized
into multiple
packages
within the
src folder.

This item includes the android.jar file that your application builds against. The version of this file was determined by the build target that you chose in the New Android Project wizard. Expanding the Android 3.1 item in the project exposes the android.jar file and its installation path, as shown in Figure 5-23. The android.jar file is the android framework.

Figure 5-23:
The Android
3.0 version
of the
android.jar
file with its
location.

You may notice that the SDK is installed in the `/Library/android/` folder, which illustrates the fact that you don't have to install the SDK in any given location. It can go anywhere in your file system (on Windows, Mac OS X, or Linux).

Assets (assets) folder

The `assets` folder is empty by default. This folder is used to store raw asset files.

A *raw asset file* could be one of many assets you may need for your application to work. A great example would be a file that contains data in a proprietary format for consumption on the device. Android has the Asset Manager class, which can return all the assets currently in the `assets` directory. Upon reading an asset, your application would consume the data in the file. If you were to create an application that had its own dictionary for word lookups (for autocomplete, perhaps), you may want to bundle the dictionary into the project by placing the dictionary file (usually, an XML or binary file such as a SQLite database) in the `assets` directory.

Android treats assets as a raw approach to resource management (think bits and bytes). You aren't limited in what you can place in the `assets` directory. Note, however, that working with assets can be a little more tedious than working with resources (see the next section), because with assets you're required to work with streams of bytes and to convert them to the objects you're after — audio, video, text, and so on. For detailed information on assets, please see the Android documentation here: `http://d.android.com/guide/topics/resources/index.html`.

Assets don't receive resource IDs like resources in the `res` directory. You have to work with bits, bytes, and streams manually to access the contents.

Resources (res) folder

The `res` folder contains the various resources for your application. Always externalize any resources (such as strings and images) that your application needs. As an example, instead of placing strings inside your code, you can create a string resource and reference that resource from within the code. (I show you how to do this in Chapter 9.) Group such resources in the `res` subdirectory that suits them best.

Make sure to provide alternative resources for specific device configurations by grouping them in specifically named resource directories. At run time, Android determines which configuration the application is running in and chooses the appropriate resource (or resource folder) from which to pull its resources. By structuring your resources this way, you can provide different user interface (UI) layouts for different users' screen sizes, or display different strings according to the user's language setting, for example.

After you externalize your resources, you can access them in code through resource IDs that are generated by the ADT in the R class (see "The mysterious gen folder," later in this chapter.

You should place each resource in a specific subdirectory of your project's res directory. The subdirectories listed in Table 5-1 are the most common types of resource folders under the parent res directory.

Table 5-1   Supported Subdirectories of the res Directory

Directory	Resource Type
anim/	XML files that define animations.
color/	XML files that define a list of colors.
drawable/	Bitmap files (.png, .9.png, .jpg, .gif) or XML files that are compiled into the following drawable resources.
drawable-hdpi/	Drawables for high-resolution screens. The hdpi qualifier stands for high-density screens. This is the same as the drawable/ resource folder except that all bitmap or XML files stored here are compiled into high-resolution drawable resources.
drawable-ldpi/	Drawables for low-resolution screens. The ldpi qualifier stands for low-density screens. This is the same as the drawable/ resource folder except that all bitmap or XML files stored here are compiled into low-resolution drawable resources.
drawable-mdpi/	Drawables for medium-resolution screens. The mdpi qualifier stands for medium-density screens. This is the same as the drawable/ resource folder except that all bitmap or XML files stored here are compiled into medium-resolution drawable resources.
layout/	XML files that define a user interface layout.
menu/	XML files that represent application menus.
raw/	Arbitrary files to save in their raw form. Files in this directory aren't compressed by the system. These are commonly used files that are of a binary type. For example, audio clips that you might play when the user taps a button or swipes the screen.

Directory	Resource Type
`values/`	XML files that contain simple values, such as strings, integers, and colors. Whereas XML resource files in other `res/` folders define a single resource based on the XML filenames, files in the `values/` directory define multiple resources for various uses. There are a few filename conventions for the resources you can create in this directory:
	* `arrays.xml` for resource arrays (storing like items together such as strings or integers).
	* `colors.xml` for resources that define color values. Accessed via the `R.colors` class.
	* `dimens.xml` for resources that define dimension values. For example, 20px equates to 20 pixels. Accessed via the `R.dimens` class.
	* `strings.xml` for string values. Accessed via the `R.strings` class.
	* `styles.xml` for resources that represent styles. A style is similar to a Cascading Style Sheet in HTML. You can define many different styles and have them inherit from each other. Accessed via the `R.styles` class.

Never save resource files directly in the `res` directory — always place them in a subdirectory. If you fail to do this, a compiler error occurs.

The resources that you save in the resource folders listed in Table 5-1 are known as *default resources* — that is, they define the default design and layout of your Android application. Different types of Android-powered devices may need different resources, however. If you have a device with a larger-than-normal screen, for example, you need to provide alternative layout resources to account for the difference.

The `resource/` mechanism inside Android is very powerful, and I could easily write a small book on it alone, but I'm going to cover only the basics in this book to get you up and running. The `resource/` mechanism can help with internationalization (enabling your app for different languages and countries), device size and density, and even resources to determine the orientation that the tablet may be in. If you want to dive into the ocean that is resources, you can find out more about them by reviewing the "Providing Resources" section in the Dev Guide of the Android documentation, located at `http://d.android.com/guide/topics/resources/providing-resources.html`.

Bin, Libs, and Referenced Libraries folders

Did I say ribs? No! I said *libs,* as in libraries. Even though this folder (the `libs` directory) isn't shown in the display of your Hello Android application's file structure, you need to be aware of it anyway. The `libs` directory contains private libraries and isn't created by default. If you need it, you have to create it manually by right-clicking the project in the Package Explorer and choosing Folder from the context menu. Eclipse asks you for the name of the folder and the name of the parent folder. Choose Hello Android, type **libs**, and click Finish.

The private libraries storied in the `libs/` directory usually are third-party libraries that perform some function for you. A good example is jTwitter, a third-party Java library for the Twitter API. If you were to use jTwitter in your Android application, you would need to place the `jtwitter.jar` library in your `libs` directory.

After a library is placed in the `libs` directory, you need to add it to your Java build path — the class path that's used for building a Java project. If your project depends on another third-party or private library, Eclipse should know where to find that library, and setting the build path through Eclipse does exactly that. For example, you can add the `jtwitter.jar` library to your build path easily by right-clicking the `jtwitter.jar` file and choosing Build Path⇨Add to Build Path from the context menu. When you do this, the `Referenced Libraries` folder, as shown in Figure 5-24, is created. This folder exists to let you know what libraries you have referenced in your Eclipse project.

You can find out more about jTwitter at `www.winterwell.com/software/jtwitter.php`.

Figure 5-24:
The Refer-
enced
Libraries
folder with
jtwitter.jar.

I don't use the `libs` directory in an examples in this book, but I discuss it here because developers — myself included — commonly use third-party libraries in Android applications. This information may be useful if you ever need to reference a library in your own Android project.

The mysterious gen folder

Ah, you finally get to witness the magic that is the `gen` folder. When you create your Android application, before the first compilation, the `gen` folder doesn't exist. Upon the first compilation, ADT generates the `gen` folder and its contents.

The `gen` folder contains Java files generated by ADT. The ADT creates an `R.java` file (more about this in a moment). I cover the `res` folder before the `gen` folder because the `gen` folder contains items that are generated from the `res` directory. Without a proper understanding of what the `res` folder is and what it contains, you have no clue how to use the `gen` folder.

Because you've externalized the resources in your Java code, you will eventually need to reference these items in the `res` folder. You do this by using the R class. The `R.java` file is an index to all the resources defined in your `res` folder. You use this class as a shorthand way to reference resources you've included in your project. This approach is particularly useful with Eclipse, whose code-completion features allow you to quickly identify the proper resource.

Expand the `gen` folder in the Hello Android project and the package name contained within the `gen` folder. Now open the `R.java` file by double-clicking it. You can see a Java class that contains nested Java classes. These nested Java classes have the same names as some of the `res` folders defined in the preceding `res` section. Under each of those subclasses, you can see members that have the same names as the resources in their respective `res` folders (excluding their file extensions). The code in your Hello Android project's `R.java` file should look similar to this:

```
/* AUTO-GENERATED FILE.  DO NOT MODIFY.
 *
 * This class was automatically generated by the
 * aapt tool from the resource data it found.  It
 * should not be modified by hand.
 */

package com.dummies.android.helloandroid;

public final class R {
    public static final class attr {
    }
```

```
public static final class drawable {
    public static final int icon=0x7f020000;
}
public static final class layout {
    public static final int main=0x7f030000;
}
public static final class string {
    public static final int app_name=0x7f040001;
    public static final int hello=0x7f040000;
}
}
```

Whoa, what's all that `0x` stuff? I'm happy to tell you that you don't need to worry about it. The ADT tool generates this code for you so that you don't need to think about what's happening behind the scenes. As you add resources and the project is rebuilt, ADT regenerates the `R.java` file to contain members that reference your recently added resources.

Never edit the `R.java` file by hand. If you do, your application may not compile, and then you're in a whole world of hurt. If you accidentally edit the `R.java` file and can't undo your changes, just delete the `gen` folder and build your project by choosing Project➪Clean. At this point, ADT regenerates the `R.java` file for you.

Viewing the application's manifest file

You keep track of everything you own and need through lists, don't you? Well, that's exactly what the Android manifest file does. It keeps track of everything your application needs, requests, and uses to run.

The Android manifest file is stored at the root of your project and is named `AndroidManifest.xml`. Every application must have an Android manifest file in its root directory.

The application manifest file provides all the essential information to the Android system — information that it must have before it can run any of your application's code. The application manifest file also provides the following:

- ✔ The name of your Java package for the application, which is the unique identifier for your application in the Android system as well as in the Android Market
- ✔ The components of the application, such as the activities and background services
- ✔ The declaration of the permissions your application requires to run
- ✔ The minimum level of the Android API that the application requires

The Android manifest file declares the version of your application. You *must* version your application. How you version your application is very similar to how the Android OS is versioned. Determining your application's versioning strategy early in the development process, including considerations for future releases of your application, is important. The versioning requirements are that each application have a version code and version name. I cover these values in the following sections. I cover versioning in more detail in Chapter 10.

Version code

The *version code* is an integer value that represents the version of the application source code. This value helps other applications determine their compatibility with your application. Also, the Android Market uses the version code as a basis for identifying the application internally and for handling updates.

You can set the version code to any integer value you like, but you should make sure that each successive release of your application has a version code greater than the previous one. The Android system doesn't enforce this rule; it's a best practice to follow.

Typically, on your first release, you set your version code to 1. Then you monotonically increase the value in a given order with each release, whether the release is major or minor. This means that the version code doesn't have a strong resemblance to the application release version that's visible to the user, which is the version name (see the next section). The version code typically isn't displayed to users in applications.

Upgrading your application source code and releasing the app without incrementing your version code causes different code bases of your app to be distributed under the same version. Consider a scenario in which you release your application with version code 1. This is your first release. A user installs your application through the Android Market and notices a bug in your application, and she lets you know. You fix the bug in the code, recompile, and release the new code base without updating the version code in the Android manifest file. At this point, the Android Market doesn't know that anything has changed because it's inspecting your version code in the application manifest. If the version code had changed to a value greater than 1, such as 2, the Market would recognize that an update had been made and would inform users who installed the version-code 1 app that an update is available. If you didn't update the version code, users would never get the update to your code base, and they would be running a buggy app. No one likes that!

Version name

The *version name* is a string value that represents the release version of the application code as it should be shown to users. The value is a string that follows a common release-name nomenclature that describes the application version:

```
<major>.<minor>.<point>
```

An example of this release-name nomenclature is 2.1.4 or, without the `<point>` value (4, in this case), 2.1.

The Android system doesn't use this value for any purpose. This value is only used to help users distinguish between versions.

The version name may be any other type of absolute or relative version identifier. The Foursquare application, for example, uses a version-naming scheme that corresponds to the date. An example of the version application name is `2011-06-08`, which clearly represents a date. The version name is left up to you. You should plan ahead and make sure that your versioning strategy makes sense to you and your users.

Permissions

Assume that your application needs to access the Internet to retrieve some data. Android restricts Internet access by default. For your application to have access to the Internet, you need to ask for it.

In the application manifest file, you must define the permissions your application needs to operate. Table 5-2 lists some of the most commonly requested permissions. Permissions are covered in more detail in Chapter 15.

Table 5-2   Commonly Requested Application Permissions

Permission	Description
Internet	The application needs access to the Internet.
Write External Storage	The application needs to write data to the Secure Digital Card (SD Card).
Camera	The application needs access to the camera.
Access Fine Location	The application needs access to the Global Positioning System (GPS) location.
Read Phone State	The application needs to access the state of the phone (such as ringing).

Viewing the default.properties file

The `default.properties` file is used in conjunction with ADT and Eclipse. It contains project settings such as the build target. This file is integral to the project, so don't lose it!

The `default.properties` file should never be edited manually. To edit the contents of the file, use the editor in Eclipse. You can access this editor by right-clicking the project name in the Package Explorer and choosing Properties from the context menu. Doing this opens the Properties editor, shown in Figure 5-25.

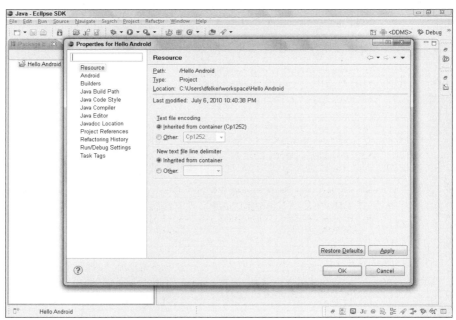

Figure 5-25:
The Properties editor in Eclipse.

This editor allows you to change various properties of the project by selecting any of the options on the left. You could select the Android property and change the path to the Android SDK, for example.

Part II

Building and Publishing Your First Android Tablet Application

The 5th Wave   By Rich Tennant

"Okay, have we all signed in on our Android Tablets? Good. I see we have Barge, Teabag, Dink, and Boob with us today."

In this part . . .

In Part II, I walk you through developing a useful Android application. I start with the basics of the Android tools and then delve into developing the screens and home-screen widgets that users will interact with. When the application is complete, I demonstrate how to sign your application digitally so that you can deploy it to the Android Market. I finish the part with an in-depth view of publishing your application to the Android Market.

Chapter 6

Designing the User Interface

Congratulations! You discovered what Android is and how to build your first application. I'm happy to say that you are now getting into the fun stuff. Over the course of the next few chapters I help you build a real application that you can use and publish to the Android Market.

What kind of application? Well, imagine you're at home, sitting at your favorite breakfast nook and reading the daily news on your tablet. The sun is shining, but you can read your tablet easily. Later that night, when you're sitting up in bed with the lights off, reading your favorite book on that same tablet isn't so easy. In fact, every time you read in bed, you have to open up the tablet's display settings and reduce the brightness to a tolerable level. Although this isn't a huge problem, it is kind of a nuisance. (Some tablets, such as the XOOM, come with features that adjust screen brightness automatically, but for the sake of this example, assume that your tablet does not offer that feature.)

It would be great if you had an application that would allow you to touch a button to switch the brightness of your display. The button would have a bright mode for daytime and another, darker mode for the evening. Tapping the button on the Home screen, then, would toggle between these settings, which would save you a lot of hassle. You'd never have to adjust the brightness again!

That's the application you're about to build. In this chapter you create the user interface portion of the application.

Creating the Screen Brightness Toggle

Your task at hand is to create the Screen Brightness Toggle application, and because you're already an expert on setting up new Android applications, I'm not going to walk you through it step by step. If you need a brief refresher on how to create a new Android app in Eclipse, review Chapter 5.

If you still have files open from the Hello Android project in Chapter 5, you need to do some tidying up before starting on this new project. First, close all the files you already have open in Eclipse. You can do this by closing each file individually or by right-clicking the files and choosing Close All from the shortcut menu.

After you have closed all the files, you need to close the current project (Hello Android) in which you're working. In Eclipse, in the Package Explorer, right-click the Hello Android project and choose Close Project. By closing the project, you are telling Eclipse that you currently do not need to work with that project. Doing this frees resources that Eclipse uses to track the project state, and therefore speeds up your application.

You're now ready to create your new Screen Brightness Toggle application. Follow these steps to get started:

1. **Create the new application by choosing File⇨New Project. Choose Android Project from the list and then click the Next button.**

2. **Fill out the project settings. For the appropriate values, refer to Table 6-1.**

Table 6-1   Project Settings for Screen Brightness Toggle

Setting	Value
Application Name	Screen Brightness Toggle
Project name	Screen Brightness Toggle
Contents	Leave the default check box (Create New Project in Workspace) selected
Build target	Android 3.1
Package name	`com.dummies.android.` `screenbrightnesstoggle`
Create activity	`MainActivity`
Min SDK Version	11

Notice how you selected the build target of Android 3.1 and a Min SDK Version of 11. By making this selection, you told Android that your code can run on any device that is running at least a version code of 11 (Android 3.0). If you were to change this to version code 8, you would be saying that your app can run on any device running version 8 or higher. How do I know which version this app can run on? I tested it, before I wrote the book! When creating a new application, you should check to see whether it can run on older versions.

3. **Click the Finish button.**

   Your new Screen Brightness Toggle application appears in your Package Explorer, as shown in Figure 6-1.

However, if you receive a "The project cannot be built until build path errors are resolved" error, you can resolve it by right-clicking on the project and choosing Android Tools⇨Fix Project Properties. Doing this realigns your project with the IDE workspace.

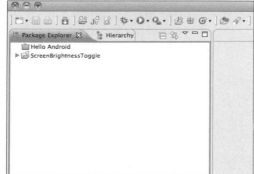

Figure 6-1:
The Screen Brightness Toggle application in Eclipse.

Understanding Layouts

Now that you have created the Screen Brightness Toggle application inside Eclipse, you can begin to design the application's user interface. Because the *user interface* is the part of your application with which your users interact, you must make this area of the application as quick and responsive as possible in all regards.

For the Screen Brightness Toggle application, all you need is a single button centered in the middle of the screen to toggle the screen brightness. Directly below the button you can include some text that lets the user know whether the tablet is in regular daytime mode or in night mode. (See Figures 6-2 and 6-3, respectively.)

Figure 6-2:
The Screen
Brightness
Toggle
application
in daytime
mode (100%
screen
brightness).

Figure 6-3:
The Screen
Brightness
Toggle
application
in night
mode (20%
screen
brightness).

Understanding the XML layout file

All layout files for your application are stored in the `res/layouts` directory of your Android project in Eclipse. Android Development Tools (ADT) creates a default layout file when you create a new application. When you created the Screen Brightness Toggle application, the ADT created a file in the `res/layouts` directory named `main.xml`.

Open that file by double-clicking it, and the XML code appears in the Eclipse editor window, as shown in Figure 6-4.

Layout files go here   The main.xml layout file

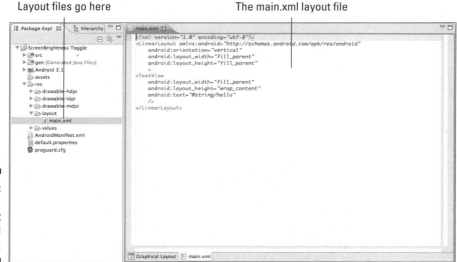

Figure 6-4:
The `main.xml` layout file opened in Eclipse.

Your XML code is a simple layout in which you have a text value in the middle of the screen, and it should look like this:

```
<?xml version="1.0" encoding="utf-8"?>                                    →1
<LinearLayout xmlns:android="http://schemas.android.com/apk/res/android"  →2
    android:orientation="vertical"
    android:layout_width="fill_parent"
    android:layout_height="fill_parent"
    >                                                                     →6
<TextView                                                                 →7
    android:layout_width="fill_parent"
    android:layout_height="wrap_content"
    android:text="@string/hello"
    />                                                                    →11
</LinearLayout>                                                           →12
```

This XML file defines exactly what your view is to look like. In the following list, I break this file down for you element by element:

→**1**   The first element provides the default XML declaration, letting text editors such as Eclipse and consumers like Android know what type of file it is.

→**2** to →**6**   The next element defines the layout type. In this case, you're working with a LinearLayout. I give you more info about LinearLayout layout types in a moment, but for now, be aware that a LinearLayout is a container for items — known as *views* — that show up on the screen. Notice how the closing </LinearLayout> tag isn't included here. The reason is that the LinearLayout tag is a container for other items. The close tag is inserted only after all the view items have been added to the container.

→**7** to →**11**   These lines define the items within the LinearLayout container. The *views* are the basic building blocks for user interface components. Lines 7–11 define TextView, which is responsible for displaying text to the screen.

A view occupies a rectangular space on the screen and is responsible for drawing and event handling. All items that can show up on a device's screen are views. The View class is the superclass that all items inherit from in Android.

→**12**   At the end of it all, you have the closing tag for the LinearLayout. This closes the container.

Arranging components with Android SDK's layout tools

When creating user interfaces, you have to lay components out in a number of different ways. Thankfully the engineering geniuses at Google who created Android thought of all this and provided you with the tools necessary to create many different types of layouts. Table 6-2 gives you a brief introduction to the common types of XML layout objects that are available in the Android Software Development Kit (SDK).

Table 6-2	Android SDK Layouts
Layout	*Description*
LinearLayout	A layout that arranges its components in a single row.

Layout	Description
RelativeLayout	A layout where the positions of the components can be described in relation to each other or to the parent.
FrameLayout	A layout designed to reserve an area on the screen to display a single item. You can add multiple components to a FrameLayout, but they will all appear pegged to the upper left of the screen, drawn in a stack, with the most recently added component at the top of the stack. This layout is commonly used as a way to lay out views in an absolute position.
TableLayout	A layout that arranges its components into rows and columns.

Other different types of layout tools exist, such as TabHost for creating tabs and Sliding Drawer for you to hide and display views with finger-swiping motions. I'm not going to get into those at this point because they are only used in special-case scenarios. The items in Table 6-2 outline the layouts that you will use most commonly.

Using the visual designer

To lay out your application, you can code Android's layouts by hand, but doing so can be complicated and difficult. However, I have some good news for you: Eclipse includes a visual designer, which allows you to drag and drop the elements of your app as you please. I also have some bad news: The designer (like all visual designers) is limited in what it can do.

The visual designer works well for simple scenarios in which your contents won't need to change. It is best suited for *static content scenarios* — that is, scenarios in which your layout is created once and is not updated dynamically. In such scenarios, the text of TextViews or images might change, but the actual layout of the views inside the layout would not change. But what happens when you need to draw items on the screen dynamically based on user input? That's where the designer falls down. It cannot help you in those scenarios.

To view the visual designer, with the main.xml file open in the Eclipse editor, click the Graphical Layout tab (see Figure 6-5).

Figure 6-5:
The Layout
button,
which
shows
the visual
designer.

Access the visual designer here

You should now see the visual designer, as shown in Figure 6-6. From here, you can drag and drop items from the Layouts or Views toolboxes.

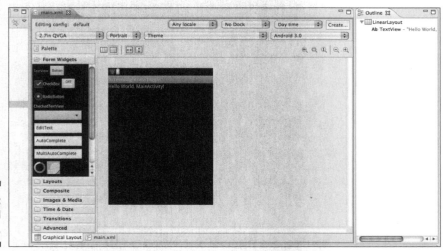

Figure 6-6:
The visual
designer.

A feature I really like about the designer is the ability to view the properties of a given view. Right-click on a view and hover your cursor over Properties to access the Properties view in Eclipse, as shown in Figure 6-7. To use the

Properties pop-up menu, simply select a view in the visual designer and right-click on it. The view has a red border around it, and the properties show up in the Properties window below. Scroll through the list of properties to examine what can be changed in the view.

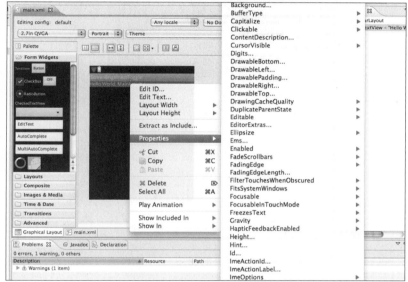

Figure 6-7:
The properties of a selected item in the visual designer.

A view's available properties can change depending on its parent layout. For example, a TextView that is inside a LinearLayout will have a different set of layout properties than one inside a RelativeLayout.

Developing the User Interface

Okay, it's time to start developing your user interface.

As stated previously, views in Android are the basic building blocks for user interface components. Anytime you implement a user interface component, such as a Layout, TextView, and so on, in the Android system, you are using a view.

So that Android knows how to lay out your view on the screen, however, you must configure a couple of attributes. These attributes, known in the Android SDK as LayoutParams, are

✔ The `layout_width` attribute, which specifies the view's width

✔ The `layout_height` attribute, which specifies the height

If you're using a static layout, these two attributes must be set in the XML layout. If you're creating views dynamically through code, the layout parameters must be set through Java code. Either way, you cannot do without them. I do not cover dynamic creation of views in this book. For more about dynamic creation of views, refer to the API samples that come with the Android SDK.

If you forget to provide values for `layout_width` or `layout_height`, your Android application will crash when rendering the view. Thankfully, you are reminded of this sort of omission quickly when you test your application.

These attributes accept any pixel value or density-independent pixel value (density-independent pixels will be explained in Chapter 9) to specify their respective dimensions. However, two of the most common values for `layout_width` and `layout_height` don't specify pixel values at all. They are

✔ The `fill_parent` value, which instructs the Android system to fill as much space as possible on the screen based on the available space of the parent layout.

As of Android 2.2, `fill_parent` has been renamed to `match_parent`. However, to maintain backward compatibility, `fill_parent` is still supported. If you plan on developing for Android 2.2 and above, use `match_parent`. If you need to support devices prior to Android 2.2, then use `fill_parent`.

✔ The `wrap_content` value, which instructs the Android system to take up only as much space as needed to show the view. As the view's contents grow, as would happen with a `TextView`, the view's viewable space grows. This feature is similar to the `Autosize` property in Windows forms development.

To define these attributes in your Screen Brightness Toggle app, first return to the XML view of your layout by clicking the `main.xml` tab, which is located directly next to the Graphical Layout tab you clicked to get to the visual designer. When you are in the XML view, replace the contents of the view — that is, all the XML code within the `main.xml` file — with the following:

```
<?xml version="1.0" encoding="utf-8"?>
<RelativeLayout xmlns:android="http://schemas.android.com/apk/res/android"   ➞2
    android:layout_width="match_parent"                                       ➞3
    android:layout_height="match_parent"                                      ➞4
    >

</RelativeLayout>
```

Let me explain the Android layout XML that you just added into your file:

→**2**   `xmlns:android="..."` defines the XML namespace that you will use to reference part of the Android SDK.

→**3**   This line is the first attribute definition. This line informs the view that it should fill as much horizontal space as it can, up to its parent. In short, it should make the width as wide as it can be within the parent.

→**4**   Similarly, this line informs the view that it should fill as much vertical space as it can, up to its parent — it should make the height as tall as it can be within the parent.

At this point, you have defined your layout to fill the entire screen by setting both the width and height to `"match_parent"`.

Adding Widgets to Your Layout

A widget is a `View` object that serves as an interface for interaction with the user. Android devices come fully equipped with various widgets, including buttons, check boxes, and text-entry fields, so that you can quickly build your user interface. Some widgets are more complex, such as a date picker, a clock, and zoom controls.

Widgets also provide user interface events that inform you when a user has interacted with the particular widget, such as tapping a button.

The Android documentation can get a bit sticky at times, and widgets and app widgets are regularly confused. They are two completely different topics. I am currently referring to widgets in the sense that you can find defined at `http://d.android.com/reference/android/widget/package-summary.html`. Please note, these are not the same as app widgets, which I will cover in Chapter 8.

For your Screen Brightness Toggle app, you need to add a single widget: the `ToggleButton` widget, which lets the user toggle the screen brightness of the tablet.

To add a `ToggleButton` to your layout, type the following code into your `main.xml` file, just before the `</RelativeLayout>` line:

```
<ToggleButton
    android:id="@+id/toggleButton"                        →2
    android:layout_width="wrap_content"                   →3
    android:layout_height="wrap_content"                  →4
    android:layout_centerInParent="true"                  →5
    android:textOn="Dimmer On"                            →6
    android:textOff="Dimmer Off"/>                        →7
```

The `ToggleButton` contains some new parameters:

→**2**   The `id` attribute defines the unique identifier for the view in the Android system. Here you inform Android that the identifier for your widget is `toggleButton`, which means you can refer to it by that name in the Java code later. Nothing beats the actual Android documentation on the subject, which is located at `http://developer.android.com/guide/topics/ui/declaring-layout.html`.

→**3** to →**4**   The height and width are set to `wrap_content`, which informs the Android layout system to place the widget on the screen and only take up as much usable space as it needs.

→**5**   Because you've placed the `ToggleButton` code within the `RelativeLayout`, the `layout_centerInParent` attribute gets inherited from `RelativeLayout`. This attribute instructs the Android system to position the `ToggleButton` within the center of its parent (that is, within the `RelativeLayout`).

→**6**   This attribute defines the `ToggleButton`'s text value when the `ToggleButton` is in its `On` state.

→**7**   Similarly, this attribute defines the `ToggleButton`'s text value when the `ToggleButton` is in its `Off` state.

You have now added a button to your view with an ID resource of `toggle-Button`. The ID is how you will reference the button in the Java code (which I get to in Chapter 8).

Adding Visual Queues for the User

Now that you have added the `ToggleButton` widget, you need to add a couple of visual queues for your end users. These queues are going to entail some instructional content as well as some larger text that will display the current percentage level of the brightness (between 1 and 100).

To add these queues, you must add two `TextView` elements in the `main.xml` file. The first, placed just above the `ToggleButton`, is as follows:

```
<TextView
    android:layout_width="wrap_content"
    android:layout_height="wrap_content"
    android:text=" Press the toggle button to toggle the screen brightness
        between 20 and 100% (night and day time)"                          →4
    android:layout_centerHorizontal="true"/>                                →5
```

The parameters in this `TextView` include

→**4**   The `android:text="..."` property sets the text of the `TextView`.

→**5**   This property instructs Android to lay out the `TextView` in the center of the screen, horizontally. Without this instruction, this `TextView`, because it's included as the first view within the `RelativeLayout`, would anchor itself at the top-left position of the screen.

The second `TextView` should be placed below the ToggleButton (and before the `</RelativeLayout>` line), as follows:

```
<TextView
    android:text="TextView"
    android:layout_width="wrap_content"
    android:layout_height="wrap_content"
    android:id="@+id/currentBrightness"
    android:textSize="55px"
    android:layout_below="@id/toggleButton"
    android:layout_centerHorizontal="true"/>
```

→**6**
→**7**

The parameters in this `TextView` include

→**6**   This property instructs Android to set the font sizes for the `TextView` to 55px.

→**7**   This property instructs Android to lay out this `TextView` below the `ToggleButton`.

Previewing the Application in the Visual Designer

That's it! You've created your layout for this application!

The full code for your layout is as follows:

```
<?xml version="1.0" encoding="utf-8"?>
<RelativeLayout xmlns:android=
    "http://schemas.android.com/apk/res/android"
    android:layout_width="match_parent"
    android:layout_height="match_parent"
    >
    <TextView
        android:layout_width="wrap_content"
```

```
          android:layout_height="wrap_content"
          android:text=" Press the toggle button to toggle the screen brightness
                  between 20 and 100% (night and day time)"
          android:layout_centerHorizontal="true"/>
    <ToggleButton
          android:id="@+id/toggleButton"
          android:layout_width="wrap_content"
          android:layout_height="wrap_content"
          android:layout_centerInParent="true"
          android:textOn="Dimmer On"
          android:textOff="Dimmer Off"/>
    <TextView
          android:text="TextView"
          android:layout_width="wrap_content"
          android:layout_height="wrap_content"
          android:id="@+id/currentBrightness"
          android:textSize="55px"
          android:layout_below="@id/toggleButton"
          android:layout_centerHorizontal="true"/>
</RelativeLayout>
```

Now you can see what the layout looks like in the visual designer. Click the Graphical Layout tab to view the visual designer, as shown in Figure 6-8.

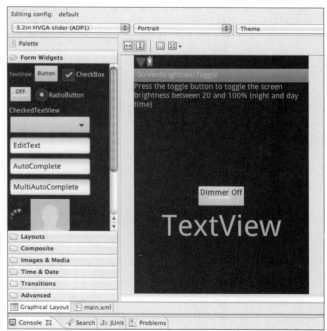

Figure 6-8:
The visual
designer
view of the
layout.

Depending on your layout, you may from time to time need to change the orientation of the visual designer from landscape to portrait (or back again). To do this, just click the Config drop-down list and select Portrait or Landscape as needed.

The ADT Visual Designer

The visual designer has many different layout configurations. By default, the designer is set to Android Development Phone 1 (ADP1). The ADP1 was the first development phone offered by Google. Developers could purchase an ADP1 device on which to test their apps. Since then, Google has released several other versions of the development phone. The Devices drop-down list in the visual designer shows you which ones you can work with. The configurations represent the various configurations that the device can be in. For example, the ADP1 had three states that were valid at run time:

✔ **Landscape, closed:** Phone is in landscape mode, physical keyboard is hidden.

✔ **Portrait:** Phone is held in portrait mode.

✔ **Landscape, open:** Phone is in landscape mode, physical keyboard is extended.

Each device in the Devices drop-down list has its own set of configurations. You can create your own custom configurations by choosing Devices⇨Custom⇨Custom⇨New.

For purposes of this book, I use the 10.1in WXGA (tablet) configuration.

Chapter 7

Coding Your Application

I'm sure that you are champing at the bit to start coding your application; I know I would be if I were you! In this chapter, you're going to be coding the logic of your application by writing Java code. But before you can start banging out some bits and bytes, you need a firm understanding of activities.

Understanding Activities

In Android, the building blocks of applications are *activities,* so before you start banging out bits and bytes, you need a firm understanding of what activities are and how they work. An *activity* is a single, focused thing a user can do or interact with in an application. For example, when an app presents a list of menu items the user can choose from, that's an activity. When an app displays photographs along with their captions, that's also an activity. An application may consist of just one activity, or like most applications in the Android system, it may contain several. Though activities may work together to appear to be one cohesive application, they are actually independent of each other. An *activity* in Android is an important part of an application's overall life cycle, and the way the activities are launched and put together is a fundamental part of the Android's application model.

In Android, each activity is implemented as an implementation of the `Activity` base class. Almost all activities interact with the user, so the `Activity` class takes care of creating the window for you in which you can place your user interface (UI). Activities are most often presented in full-screen mode, but in some instances, you can find activities floating in windows or embedded inside other activities — this is known as an *activity group*.

Working with stacks and states

Activities in the system are managed as an *activity stack*. As users work with a tablet, they move from activity to activity, sometimes across applications. When a new activity is created, it is placed on top of the stack and becomes the running activity. The previous running activity always remains below it in the stack and does not come to the foreground again until the new activity exits.

I cannot stress enough the importance of understanding how and why the activity works behind the scenes. Not only will you come away with a better understanding of the Android platform, but you will also be able to accurately troubleshoot why your application is doing very odd things at run time.

An activity has essentially four states, as shown in Table 7-1.

Table 7-1	Four Essential States of an Activity
Activity State	*Description*
Active/running	The activity is in the foreground of the screen (at the top of the stack).
Paused	The activity has lost focus but is still visible (that is, a new non-full-sized or transparent activity has focus on top of your activity). A paused activity is completely alive, meaning that it can completely maintain state and member information and remains attached to the window manager that controls the windows in Android. However, note that the activity can be killed by the Android system in extreme low-memory conditions.
Stopped	If an activity becomes completely obscured by another activity, it is stopped. It retains all state and member information, but it is not visible to the user. Therefore, the window is hidden and will often be killed by the Android system when memory is needed elsewhere.
Create and resuming	The system has either paused or stopped the activity. The system can either reclaim the memory by asking it to finish or it can kill the process. When it displays that activity to the user, it must resume by restarting and restoring to its previous state.

Tracking an activity's life cycle

Pictures are worth a thousand words, and, in my opinion, flow diagrams are worth ten times that. The following diagram (Figure 7-1) shows the important paths of an activity. This is the *activity life cycle*.

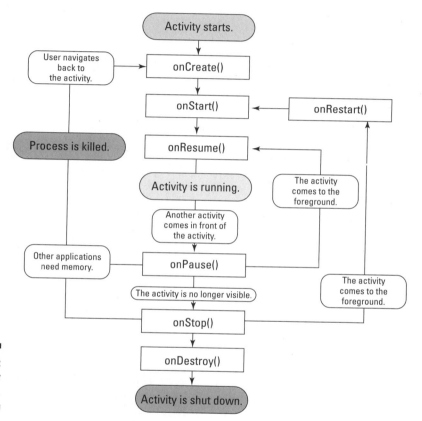

Figure 7-1:
The activity
life cycle.

The rectangles represent callback methods you can implement to respond to events in the activity. The shaded ovals are the major states that the activity can be in.

Take note of these three loops in your activity:

✔ The **entire lifetime** takes place between the first call to onCreate() and the final call to onDestroy(). The activity performs all the global setup in onCreate() and releases all remaining resources in onDestroy(). For example, if you create a thread to download a file from the Internet in the background, that may be initialized in the onCreate() method. That thread could be stopped in the onDestroy() method.

✔ The **visible lifetime** of the activity takes place between the onStart() and onStop() methods. During this time, the user can see the activity on the screen (though it may not be in the foreground interacting with the user — this can happen when the user is interacting with a dialog box). Between these two methods, you maintain the logic needed to show and run your activity. For example, say that you create an event handler to monitor the state of the phone. When the tablet state changes, this event handler informs the activity that the tablet is, say, going into Airplane mode, so the activity can react accordingly. You would set up the event handler in onStart() and tear down any resources you are accessing in onStop(). The onStart() and onStop() methods can be called multiple times as the activity becomes visible or hidden to the user.

✔ The **foreground lifetime** of the activity begins at the call to onResume() and ends at the call to onPause(). During this time, the activity is in front of all other activities and is interacting with the user. It is normal for an activity to go between onResume() and onPause() multiple times, for example, when the device goes to sleep or when a new activity handles a particular event; therefore, the code in these methods must be fairly lightweight.

The activity life cycle is a large and complex topic. I give you the basics here so that you understand the applications you will be building in this book, but I highly advise that you read through the Activity Life Cycle and Process Life Cycle portions of the Android documentation at http://d.android.com/reference/android/app/Activity.html#ActivityLifecycle and http://d.android.com/reference/android/app/Activity.html#ProcessLifecycle.

If you are interested in breaking apart your application activities into different manageable chunks, then Fragments may interest you. Fragments were introduced in Android 3.0 to allow you to build more sophisticated and modularized layouts. Fragments are a beast all unto their own and a full discussion of them requires more space than I have in this book. If you are interested, please see the Fragment documentation online at http://d.android.com/guide/topics/fundamentals/fragments.html.

Understanding activity methods

The entire activity life cycle boils down to the following methods. All methods can be overridden, and custom code can be placed in all of them. All activities implement onCreate() for initialization and may also implement `onPause()` for cleanup. You should always call the superclass (base class) when implementing these methods:

```
public class Activity extends ApplicationContext {
    protected void onCreate(Bundle savedInstanceState);
    protected void onStart();
    protected void onRestart();
    protected void onResume();
    protected void onPause();
    protected void onStop();
    protected void onDestroy();
}
```

In general, the movement an activity makes through its life cycle looks like this:

✔ `onCreate()`: Called when the activity is first created. This method is where you initialize your activity and most of your activity's class-wide variables. Most importantly, this is where you tell the activity (using layout resource identifiers) what layout to use. This method is considered the entry point of your activity. Killable: No. Next: `onStart()`.

✔ `onRestart()`: Called after your activity has been stopped prior to being started again. `onStart()` is always called next. Killable: No. Next: `onStart()`.

✔ `onStart()`: Called when your activity is becoming visible to the user. Followed by `onResume()` if the activity is brought to the foreground or `onStop()` if it becomes hidden from the user. Killable: No. Next: `onResume()` or `onStop()`.

✔ `onResume()`: Called when the activity will be available for interacting with the user. The activity is at the top of the activity stack at this point. Killable: No. Next: `onPause()`.

✔ `onPause()`: Called when the system is about to resume a previous activity or when the user has navigated away to another portion of the system, such as by pressing the home key. Any changes to data made by the user should be committed at this point (that is, if you need to save them). If the activity is brought back to the foreground, `onResume()` is called; if the activity becomes invisible to the user, `onStop()` is called. Killable: Yes. Next: `onResume()` or `onStop()`.

✔ `onStop()`: Called when the activity is no longer visible to the user because another activity has resumed and is covering this one. This may happen because another activity has started or a previous activity has resumed and is now in the foreground of the activity stack. Followed by `onRestart()` if this activity is coming back to interact with the user or `onDestroy()` if this activity is going away. Killable: Yes. Next: `onRestart()` or `onDestroy()`.

✔ `onDestroy()`: The final call you receive before your activity is destroyed. This method gets called either because the activity is finishing (such as someone calling `finish()` on it or because the system is temporarily destroying the activity to reclaim space). You can distinguish between these two with the isFinishing() method, which helps identify whether the method is actually finishing or the system is killing it. The isFinishing() method is often used inside of onPause() to determine whether the activity is pausing or being destroyed. Killable: Yes. Next: Nothing.

The methods marked as killable can be "killed" by the Android system at any time without notice. Because of this, you should set up the onPause() method to do any last cleanup and to write any persistent data (such as user edits to data) to your storage mechanism.

Recognizing configuration changes

Configuration changes are important because they can affect an activity's life cycle. *A configuration change* is defined as a change in screen orientation (the user moving the screen to the side and then back, portrait to landscape and vice versa), language, input devices, and so on. When a configuration change happens, your activity's resources, layout files, and so on might need to change in order to work properly. For example, an application may look completely different if it is interacting with the user in portrait mode as compared to landscape mode (on its side). To deal with this, when a configuration change occurs, your activity is destroyed and goes through the normal activity end-of-life cycle of onPause()⇨onStop()⇨onDestroy(). After the onDestroy() method is called, the system creates a new instance of the activity, one that uses the new system configuration.

Creating Your First Activity

Now you can create your first activity! Well, technically speaking, you already created your first activity — the MainActivity.java file — when you created a project through the New Android Project wizard in Chapter 5. Therefore, you will not be *creating* a new activity; you will only be working with the MainActivity.java file you already created. Open this file now.

As you read previously, the entry point into your application is the onCreate() method. The code for your MainActivity.java file already contains an implementation of the onCreate() method. Right now, your code should look like this:

```
public class MainActivity extends Activity {
    /** Called when the activity is first created. */
    @Override
    public void onCreate(Bundle savedInstanceState) {          →4
        super.onCreate(savedInstanceState);                    →5
        setContentView(R.layout.main);                         →6
    }
}
```

You will be writing initialization code directly below the `setContentView()` method shortly.

Some points to note about this code:

→**4**   This line is the `onCreate` method, which initiates the activity. The `Bundle savedInstanceState` part ensures that upon initiation, a bundle containing this activity's previous state (if there was one) gets passed in. This is useful when restarting an activity that has been killed by the system, so information is not lost.

The *bundle* is a key value that maps between string keys and various parcelable types. A bundle gives you, the developer, a way to pass groups of information back and forth between screens (that is, between different activities). I discuss this later in Part III, when I walk you through building the Task Reminder application.

→**5**   This line calls to the base `Activity` class to perform setup work for the `MainActivity` class.

This line is required for the application to run. You must always include this method call to your `onCreate()` method or you'll receive a run-time exception.

→**6**   This line sets the *content view* for an activity — in other words, it defines the user interface layout.

By default, an activity has no idea what its UI is. UIs come in many varieties: anything from a simple user-input form to a visual camera–based augmented virtual reality application (such as Layar in the Android Market) to a drawn-on-the-fly 2D or 3D game. As a developer, your job is to tell the activity which layout to load.

`R.layout.main` is the `main.xml` file that is located in the `res/layouts` directory. You define this layout in the previous chapter.

Handling user input

For user interaction, your Screen Brightness Toggle application won't require much: only a single button. The user taps the button to dim the tablet's screen brightness, and then the user taps the button again to restore it to full brightness.

To respond to this tap event, you need to register what is known as an event listener. An *event listener* detects and responds to an event in the Android system. Android uses various types of events, but two of the most common are

✔ **Keyboard events:** These occur when a particular keyboard key has been pressed. Why would you want to know about this? Maybe you want your app to use *hot keys,* such as Alt+E, to change your app on the fly. By pressing Alt+E, then, the user could, say, toggle an app's view into Edit mode. Responding to keyboard events allows you to do this. I do not use keyboard events in this book, but if you need to in future applications, override the onKeyDown method, as shown here:

```
@Override
public boolean onKeyDown(int keyCode, KeyEvent event) {
        // TODO Auto-generated method stub
        return super.onKeyDown(keyCode, event);
}
```

✔ **Touch events (also known as touches, clicks, or taps):** These occur when the user taps a widget on the screen. Examples of widgets that can respond to touch events include (but are not limited to):

   • Button

   • ImageButton

   • EditText

   • Spinner

   • List Item Row

   • Menu Item

All views in the Android system can react to a tap; however, some of the widgets have their *clickable* property set to false by default. You can override this setting in your layout file or in code to allow a view to be clickable by setting the clickable attribute on the view or the set-Clickable() method in code.

Writing your first event handler

When the user touches the onscreen toggle button, a click event occurs. In order for your application to respond to the user's action, then, you must design the code to respond to the button's click event.

Type the code shown in Listing 7-1 into your editor. This code shows how to implement a click handler for the toggleButton. The listing shows the entire code — the original onCreate() method with the new code added. Feel free to either fill in the button code or overwrite your entire onCreate code.

Listing 7-1:   The Initial Class File with a Default Button OnClickListener

```
public class MainActivity extends Activity {

    ToggleButton tb;
                                                        →3
    TextView tv;
                                                        →4


    @Override
    public void onCreate(Bundle savedInstanceState) {
        super.onCreate(savedInstanceState);
        setContentView(R.layout.main);

        tb = (ToggleButton)findViewById(R.
           id.toggleButton);                            →11
        tv = (TextView)findViewById(R.
           id.currentBrightness);                       →12

        tb.setOnClickListener(new View.OnClickListener()
          {                                             →14

           public void onClick(View v) {
                                                        →16


              }
           });
    }
}
```

Whoa! A lot of stuff is going on there! I'm going to break this code down, line by line, so you can understand what's happening.

→**3**   The `ToggleButton` instance variable `tb`.

→**4**   The `TextView` instance variable `tv`.

Lines 3 and 4 declare class-level variables so that you can keep track of your `ToggleButton` and `TextView`. You'll need these variables throughout the chapter. Declaring them here at the top means that you won't have to constantly re-declare them every time you need them.

→**11**   Retrieves a reference to the current brightness `TextView` from the activity. This is explained in detail below.

→**12**   Retrieves a reference to the `ToggleButton` that will toggle the brightness of the screen.

Lines 10 and 11 find views through the findViewById() method, which is available to all activities in Android. This method lets you find any view inside the activity's layout. This method always returns a View class that you must cast to the appropriate type before you can work with it. As an example, the following code (which is a snippet from the previous code) casts the returned view from findViewById() to a ToggleButton (which is a subclass of View):

```
tb = (ToggleButton)findViewById(R.id.toggleButton);
```

If you cast the retrieved view to a different type than the one in your layout file (if you, say, cast the view to ImageButton, but you have an ImageView in the layout file), you will receive a run-time error. Be sure you're casting to the appropriate type.

As with most languages that implement inheritance, Android also allows you to return a parent object type. As an example, if your view contains a ToggleButton, you can cast this to a Button, TextView, or plain View. Look at the inheritance chain in the documentation to see what the view inherits so that you can properly cast. (For ToggleButton, check out http://d.android.com/reference/android/widget/ToggleButton.html.)

Immediately following the code on line 11, you retrieve the TextView that you will use to display the current brightness of the device back to the user. Though not used in this example, you need this object, so add it now to save time later in this chapter.

→14 Sets up a click listener that fires when the user taps on the ToggleButton.

→16 The click event that fires when the user taps on the ToggleButton.

Line 13 creates an event handler. The event-handling code is placed inline after you retrieved the ToggleButton from the layout. Setting up the event handler is as simple as setting a new View.OnClickListener. This click listener contains an onClick() method (line 15) that is called after the button is tapped. This is where you will be placing the code to handle the screen brightness toggle.

Importing packages

When you typed this code into your editor, you may have seen some red squiggly lines, as shown in Figure 7-2. These lines are Eclipse's way of telling you, "Hey! I don't know what this 'Button' thing is." If you place your cursor over the squiggly line and leave it there for a moment, you receive a small context window that gives you several options (also shown in Figure 7-2).

Figure 7-2:
When Eclipse gives you a red squiggly line, hovering your cursor over the line provides a context menu of options.

To help Eclipse understand, choose the first option from this menu, Import 'ToggleButton'. Doing this adds the following import statement to the top of the file:

```
import android.widget.ToggleButton;
```

This `import` statement informs Eclipse where the `ToggleButton` is located in the Android packages. You may also need to import the `android.view.package`.

If you do not import the required libraries as just explained, the code will not compile and you will not be able to complete your application.

As you start to develop more applications, and to include many other widgets in your applications, you will notice that you have to include quite a few import statements to get your application to compile. Although this is not a huge issue, you can provide a shorthand method of including everything in a particular package. You can do this by providing an asterisk at the end of the package name, as shown here:

```
import android.widget.*;
```

The asterisk informs Eclipse to include all widgets in the `android.widget` package.

Working with the Android Framework Classes

You're now getting into the good stuff: the real nitty-gritty of Android development — the Android framework classes! Yes, activities, views, and widgets are integral parts of the system, but when it comes down to it, they're plumbing and they're required in any modern operating system (in one capacity or another). The real fun is just about to start.

In the following section, you add a class to your app that determines whether the device's brightness setting is currently in a daytime mode (bright) or in evening mode (dim) and that sets the brightness to the other mode. At that point, you can start toggling the device brightness.

To add these classes, you could type all of this into the `MainActivity. java` file — but then the application would start to get long and complicated. My grandparents used to remind me that the key to keeping anything clean is good organization, and I believe the same is true for code. To keep things clean, then, you are going to create another Java file in your application that will house the code for implementing the brightness toggling logic.

Retrieving and setting screen brightness

Add a new file with the name of `BrightnessHelper.java` to your `src/` folder. Inside of this class, you are going to add two static methods with the method names `setCurrentBrightness` and `getCurrentBrightness`. The basic code for the class should look like this (here I exclude the specifics of the two methods — I get to these later — and I exclude the import statements for brevity):

```
public class BrightnessHelper {

    public static void setCurrentBrightness(Activity activity, int brightness) {

    }

    public static int getCurrentBrightness(Context context) {

    }
}
```

The two methods in the `BrightnessHelper` class file are static methods because they do not need to maintain any application state within them. The contents of each method will only need to get or set the current brightness of the device.

Now that you have a sense of how this class should look, you can begin filling in the methods. Logically speaking, before you can know how you should change the brightness, you need to determine the current brightness of the device. The first method you add is the `getCurrentBrightness` method. In your `BrightnessHelper.java` class, change your `getCurrentBrightness` method to look like Listing 7-2.

Listing 7-2:   Retrieving the current brightness

```
public static int getCurrentBrightness(Context context) {
    int brightness = 0;                                          →2
    try {                                                        →3
      brightness =
             Settings.System.getInt(context.getContentResolver(),
             Settings.System.SCREEN_BRIGHTNESS);                 →6
    } catch (SettingNotFoundException e) {                       →7
        Log.e(context.getPackageName(), e.getMessage(), e);      →8
    }
    return brightness;                                           →10
}
```

Here's what these lines of code do:

→2   This line defines the `brightness` variable. When the system returns this variable, on line 4, it will contain an integer value for the brightness setting. Here you set the default value to zero so that, in a worst-case scenario (if, say, an error occurs in the next few lines of code), the screen would become brighter, rather than dimmer.

→3   The beginning of a try/catch block. A try/catch block acts as a safety mechanism for your code. When an error occurs within the try block, the catch block catches it (if you specify the exception type) and then allows you to perform some sort of action. In this instance, line 3 is the beginning of the try block and line 6 is the beginning of the catch block.

→6   This line retrieves the brightness from the Android system. You are delving into the `Setttings` object and retrieving an integer value for the setting whose identifier is `Settings.System.SCREEN_BRIGHTNESS`. When importing this class, please use the import `android.provider.Settings`.

→7   On this line the catch block is instantiated and begins seeking any exception of type `SettingNotFoundException`. If the exception is any other type of exception, the method will fail and the exception will bubble up the stack until the application is stopped (the application crashes). The reason you are only looking for the `SettingNotFoundException` is that this is the only logical exception that could be thrown here. If any other exception occurs, something strange has happened and the developer should investigate. Letting the application fail at this point allows the developer to review the logs to determine the cause.

→8   This line logs the exception caught by the catch block to the Android error log. The log helps the developer (you) debug the code. (I cover debugging later in this chapter.)

→**10**   This line returns the brightness value determined in line 4 back to the calling code at this point. If the screen brightness is 42, then a value of 42 is returned. If the value is 76, then 76 will be returned.

Setting the current screen brightness is a little more complicated, but this process is nothing that should scare you off by now (heck, you're almost a seasoned pro by this point!). Replace the setCurrentBrightness method with the code shown in Listing 7-3.

Listing 7-3:   Setting the screen brightness

```
public static void setCurrentBrightness(Activity activity,
    int brightness) {

    // Change the screen brightness mode to Manual
    Settings.System.putInt(activity.getContentResolver(),
        Settings.System.SCREEN_BRIGHTNESS_MODE,
        Settings.System.SCREEN_BRIGHTNESS_MODE_MANUAL);          →7

    // Set the screen brightness for the system.
    Settings.System.putInt(activity.getContentResolver(),
        android.provider.Settings.System.SCREEN_BRIGHTNESS,
        brightness);                                             →12

    // Adjust the current window brightness to reflect the new system brightness.
    LayoutParams params =
        activity.getWindow().getAttributes();                    →16

    // Brightness is determined from values of 1 through 255
    if(brightness == 255) {                                      →19
        params.screenBrightness = 1.0f;                          →20
    } else {
        params.screenBrightness = 0.2f;                          →22
    }

    activity.getWindow().setAttributes(params);                  →23
}
```

These lines do the following:

→**7**   Whoa! This line is doing a whole bunch of things. Here is the breakdown. The Settings.System.putInt() method requires a ContentResolver object to do its job. A ContentResolver object is a built-in Android mechanism that can access and modify data outside of an application — in this case, the settings storage mechanism (think database). In other words, the putInt() method needs a storage mechanism, and it accesses it using the activity.getContentResolver() method. (This method is passed in as a method parameter from the calling code. The code that calls this method will be your MainActivity.java file, as you see later when you tie everything together.) The putInt() method tells Android to update

the SCREEN_BRIGHTNESS_MODE setting to SCREEN_BRIGHTNESS_
MODE_MANUAL. This update allows you to manually control the
brightness, so Android's automatic screen brightness code won't
interfere with what you're doing.

ContentResolver objects are an intricate part of Android and
are part of the content provider model. For more about them, look
in the documentation here: http://d.android.com/guide/
topics/providers/content-providers.html.

→**12** This line does something very similar to line 4, except that instead
of modifying a setting's mode it modifies a setting's value. The
brightness value is passed into this call along with the name
of the setting to be modified — SCREEN_BRIGHTNESS. (The
brightness variable comes from the MainActivity.java file,
as you see later.)

Even though this line updates the system's screen brightness setting, it won't adjust the brightness level in your currently running
window. The reason is because your application, when launched,
inherits its screen brightness value from the system screen brightness setting. Your app won't track updates to the system settings,
however, and it won't reflect these changes automatically. You
have to manually update it. The next few lines do this for you.

→**16** This line retrieves the layout parameters for the window. The next
few lines of code determine whether the window's current brightness should be increased or decreased. Please use the import
from android.view.WindowManager.LayoutParams for the
import statement.

→**19** This line determines whether the system brightness is equivalent
to 255.

→**20** If the system brightness level is 255 (bright), then this line
changes the current window's screen brightness parameters to
1.0f (full brightness). The window brightness is a different value
than the system brightness. The window brightness is a value
from 0 to 1 (think percentages) while the system setting relies on
values from 0 to 255.

→**22** If the system brightness level is not 255, then this line sets the
current window's screen brightness parameter to 0.2f (20%
brightness).

→**23** This line updates the current window's layout parameters.
Because you updated the screen brightness value, the brightness
will now be changed on the device.

This class is now complete! The application will now change the system brightness as well as the brightness on the currently active application window.

Toggling screen brightness

Now that you have a class that handles getting and setting the screen brightness of the device, you can implement the logic that determines whether to make the screen bright or dim. You do this in the `MainActivity.java` file. Open that file now and replace the contents with the code shown in Listing 7-4.

Listing 7-4:   The Toggle Logic

```
public class MainActivity extends Activity {

    ToggleButton tb;
    TextView tv;

    @Override
    public void onCreate(Bundle savedInstanceState) {
        super.onCreate(savedInstanceState);
        setContentView(R.layout.main);

        tb = (ToggleButton)findViewById(R.id.toggleButton);
        tv = (TextView)findViewById(R.id.currentBrightness);

        tb.setOnClickListener(new View.OnClickListener() {

            public void onClick(View v) {
                int currentBrightness =                              →17
                BrightnessHelper.getCurrentBrightness(MainActivity.this);
                if(currentBrightness > 51) {
                    // Dim it to 20 %. 20% of 255 is 51
                    BrightnessHelper.setCurrentBrightness(MainActivity.this, 51);
                                                                     →21
                } else {
                    // Push it back to full brightness
                    BrightnessHelper.setCurrentBrightness(MainActivity.this, 255);
                                                                     →24
                }
            }
        });
    }
}
```

That's a lot of new stuff happening! Following is a brief explanation of what each new section of code does:

→**17**   This line retrieves the current brightness level so you can determine whether or not you should make the screen bright or dim.

→**21**   If the retrieved brightness value was greater than 51 (20% of the maximum brightness value of 255 is 51), this sets the current brightness to 20% — a comfortable brightness setting for evening use.

→**24** If the retrieved brightness value was not greater than 51, this line sets the screen brightness 255.

Congratulations! You've now added code to make the screen toggle back and forth between bright and dim. The last step is to let the user know that something happened!

Adding Visual Feedback

At this point, your application does what it needs to do. However, it's not very user friendly. The application is missing a few key components to make it usable:

✔ A way to inform the user that something has happened.

✔ A way to let the user know what state the device is in. Dim or bright? (That is, 20% brightness or 100% brightness?)

✔ A way to update the information about the state of the device when the button is clicked.

All of these are pretty simple to implement. You just need to update the state of the ToggleButton as "On" when the screen is dim, and "Off" when the screen is bright. You also must update the TextView directly below the ToggleButton to reflect which brightness percentage the device is in (either 20% or 100% brightness).

To do this, add the code in Listing 7-5 to the MainActivity.java file.

Listing 7-5: Providing the user with feedback through the user interface

```
public class MainActivity extends Activity {

    ToggleButton tb;
    TextView tv;

    @Override
    public void onCreate(Bundle savedInstanceState) {
        super.onCreate(savedInstanceState);
        setContentView(R.layout.main);

        tb = (ToggleButton)findViewById(R.id.toggleButton);
        tv = (TextView)findViewById(R.id.currentBrightness);

        tb.setOnClickListener(new View.OnClickListener() {

            public void onClick(View v) {
              int currentBrightness =
```

(continued)

Listing 7-5 *(continued)*

```
               BrightnessHelper.getCurrentBrightness(MainActivity.this);
               if(currentBrightness > 51) {
                   // Dim it to 20 %. 20% of 255 is 51
                 BrightnessHelper.setCurrentBrightness(MainActivity.this, 51);
                 renderUi(51);                                                →22
               } else {
                   // Push it back to full brightness
                 BrightnessHelper.setCurrentBrightness(MainActivity.this, 255);
                 renderUi(255);                                               →26
               }
             }
           });

           int initialBrightness =
               BrightnessHelper.getCurrentBrightness(this);                  →32
           renderUi(initialBrightness);                                      →33
       }

       private void renderUi(int brightness) {                              →36
         if(brightness <= 0) {
             tv.setText("Default");                                         →38
         } else {
             tv.setText((((brightness / 255.0f)*100) + "%");                →40
         }

         if(brightness == 51) {
             tb.setChecked(true);                                           →44
         } else {
             tb.setChecked(false);                                          →46
         }
       }
     }
}
```

These lines do the following:

→22   This line calls the renderUi method with the value of 51 for the brightness. (The renderUI method initializes the TextView and ToggleButton on the app to the correct state.)

→26   This method calls the renderUi method with the value of 255 for the brightness.

→32   This line gets the initial brightness of the device. This line executes when the application is started; it retrieves the brightness value so your application can set the brightness properly and display the correct value.

→33   This method calls the renderUi method with the current brightness value so that the screen can initialize properly.

→36   This line defines the renderUi method.

→38   If the brightness is less than or equal to 0, the value of "default" will be placed into the `TextView`. This may happen if your application encounters an exception when retrieving the value from the `getCurrentScreenBrightness()` method.

→40   This line sets the text of the current brightness level. This lets the user know what state the dimmer is in. To get a correct percentage, you divide the brightness value by 255 (the max brightness). Multiplying the result by 100 (to eliminate the fraction) and appending a % symbol produces a value like 20% or 100% for the `TextView`.

→44   If the brightness value is set to 51, then this line sets the `ToggleButton` attribute `checked` to true. This causes the `ToggleButton` to have an "On" state. At this point the "Dimmer On" text will be shown on the `ToggleButton`.

→46   If the brightness is 255, this line sets the `ToggleButton` attribute `checked` to false. This causes the `ToggleButton` to be in an "Off" state. At this point the "Dimmer Off" text will be shown on the `ToggleButton`.

Obtaining permission to change system settings

Before you can compile and install your application, you must update the `AndroidManifest.xml` file to include the requested application permissions that this application needs to operate. The `AndroidManifest.xml` file declares what is in your application and what permissions are required for the application to run.

Each application in Android runs within its own process space (sometimes called a *sandbox*) in the Android operating system. This allows the application to perform its work without interfering with other applications on the device. Sometimes, however, your application needs to use parts of the device that the user might be interested in knowing about. A common example would be the Internet permission (if your app needs to access the Internet for any reason). Your application will not be using the Internet, so you do not have to worry about Internet permission. However, this application is changing system settings, so you need to update the `AndroidManifest.xml` file to inform the user of this. When a user installs an application from the Android Market, he is informed of the requested permissions this application requires and can decide whether to install the application. The user does not have the option to decide that an application can have one permission over another. The Android Market does not offer that level of granularity because, if it did, your app would not work. (It would crash because it would not have sufficient permission at run time.)

To add the appropriate permission to your application, open the `Android Manifest.xml` file by double-clicking it. Select the AndroidManifest.xml tab at the bottom of the screen to see the xml definition. Your file should look similar to Listing 7-6. You need to add one line, the bold line shown in Listing 7-6.

Listing 7-6:   The Updated AndroidManifest.xml File with Permissions Added

```xml
<?xml version="1.0" encoding="utf-8"?>
<manifest xmlns:android="http://schemas.android.com/apk/res/android"
    package="com.dummies.android.screeenbrightnesstoggle"
    android:versionCode="1" android:versionName="1.0">
    <uses-permission android:name="android.permission.WRITE_SETTINGS"></uses-
            permission>
    <application android:icon="@drawable/icon" android:label="@string/app_name">
        <activity android:name=".MainActivity" android:label="@string/app_name">
            <intent-filter>
                <action android:name="android.intent.action.MAIN" />
                <category android:name="android.intent.category.LAUNCHER" />
            </intent-filter>
        </activity>
        <activity android:name=
            "com.dummies.android.screeenbrightnesstoggle.ToggleActivity"
            android:theme="@android:style/Theme.Translucent" />
    </application>
    <uses-sdk android:minSdkVersion="11" />
</manifest>
```

With this permission in place, you can run the application.

Forgetting to add permission to your application will not prevent the code from compiling. However, when the application is deployed to the end user's device it will crash when it executes the code that requires that particular permission to run.

Congratulations! You've coded your first Android application!

Installing Your Application

You've done it! You wrote your first application. In the next couple of sections, you install your app on the emulator and then get it into action on a real device!

Installing on the emulator

Installing your app on an emulator is the next step. In Chapter 5, you set up a run configuration to run the Hello Android application. Now you will be using the same launch configuration for your new app. Because the ADT is smart enough to know about this launch configuration, it will use it by default. To install this app on the emulator, follow these steps:

1. **In Eclipse, choose Run⇨Run or press Ctrl+F11 to run the application.**

   You are presented with the Run As window, as shown in Figure 7-3. Choose Android Application and click the OK button. Doing this starts the emulator.

   After the emulator is running, it's running on its own. The emulator has no dependencies on Eclipse. In fact, you can close Eclipse and still interact with the emulator.

   The emulator and Eclipse speak to each other through the Android Debugging Bridge (ADB). ADB is a tool that was installed with the Android Development Tools (ADT).

Figure 7-3: The Run As configuration dialog box.

2. **Wait for the emulator to load and then unlock the emulator.**

   When the emulator starts, the screen will be in a locked state. If you're not sure how to unlock the emulator screen, simply click and drag the circular lock icon to the right. To review more about the emulator, see Chapter 5. When the emulator is unlocked, your application should start. If it does not start, rerun the application by choosing Run⇨Run or pressing Ctrl+F11. After the application starts, you should see the emulator running your program, as shown in Figure 7-4.

   You've already performed the first step in your testing process — making sure that the app starts!

3. **Test your device. Click the button to see the text change, as shown in Figure 7-5.**

4. **Return to the Home screen by clicking the home button on the emulator.**

   Open the application (by selecting the application icon in the apps menu). Note that the application launcher icon is now present in the list of applications.

Figure 7-4:
The emulator running the application.

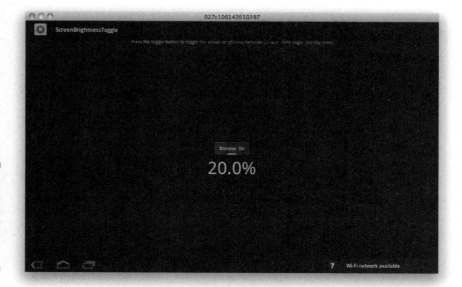

Figure 7-5:
The Screen Brightness app in Dimmer On mode.

When you opened your app, you saw the onscreen text change — but is your app actually affecting the tablet's brightness? Open the app again and check to see whether the tablet is in bright or dim mode. Aha! You've just discovered one of the drawbacks of the emulator. It can't emulate screen brightness. Why not? Your computer controls the screen brightness, not your android app. Therefore, the only way to check if this app is working is to check the brightness settings. With the app open, toggle the application to dim mode (20%). Then return to the Home screen, press Apps➪Settings, and then Screen➪Brightness. You should see a screen similar to the one in Figure 7-6.

Now go back to your app and click the button to toggle the brightness to bright mode, and then return to the brightness settings. (Yes, I know doing this is a pain, but with the emulator's limitations, it's the only way you can be sure your app is working.) Did the brightness change?

Test your application to ensure it works as expected. If you find a flaw, use the debugging tools I discuss later in this chapter to help identify and correct the problem.

Figure 7-6: The brightness scale in the screen brightness settings.

What about automated testing?

With the rise of agile methodologies over the last decade, it's only a matter of time before you start to wonder how to perform automated testing with Android. The SDK installs Android unit testing tools that you can use to test not only Java classes but also Android-based classes and user interface interactions. You can find out more about unit testing Android from the Android documentation at `http://d.android. com/guide/topics/testing/ testing_android.html`.

An entire book could easily be written on unit testing with Android alone; therefore, I'm going to mention the tools that are at your disposal. You can look into them when you have time:

✔ **jUnit:** The SDK installs jUnit integration with the ADT. jUnit is a very popular unit testing framework that is used in Java. You can use jUnit to perform unit testing or interaction testing. More info about jUnit can be found at `www.junit.org`. To make life easier, Eclipse has built-in tools to help facilitate testing in jUnit through Eclipse.

✔ **Monkey:** Monkey is a UI/application exerciser. This program runs on your emulator or device and generates pseudorandom streams of user events, including taps, gestures, touches, clicks, and a number of system events. Monkey is a great way to stress-test your application. It is installed with the Android SDK.

✔ **Robolectric:** A new type of testing framework that allows you to somewhat "mock" out various functions in the Android Framework. This library is under constant development. With each passing day, it gets better. This tool is not installed by the Android SDK. To install and for more information, go to `www.robolectric.org`.

✔ **Robotium:** As the tagline states "It's like Selenium for Android" (Selenium is an automated website-testing framework). Robotium features automated testing that is similar to Monkey's but in a much more friendly manner. This tool is not installed by the Android SDK. To install and for more information, go to `http://code. google.com/p/robotium/`.

Installing on a physical Android device

Installing the application on a device is no different than installing an application on the emulator. You just need to make a few small adjustments. You installed the driver in Chapter 4, so the rest is fairly straightforward:

1. **Enable the installation of non–Android Market applications by going to Settings⇨Applications and selecting the Unknown Sources check box.**

2. **From the Home screen of your tablet, access the Settings panel (choose Apps⇨Settings). Within that panel, choose Applications.**

3. **Select the Unknown Sources check box, as shown in Figure 7-7.**

   This setting allows non-Android market apps to be installed on the device.

Figure 7-7:
Setting to
allow the
installation
of non–
Android
Market
applica-
tions.

4. **While in the Applications Settings screen (the same screen where you made the last change), choose Development and select the USB Debugging option, as shown in Figure 7-8.**

   This setting allows you to debug your application on your device (more on debugging in a moment).

Figure 7-8:
Enabling
your device
to perform
USB
debugging.

5. **Plug your tablet into the computer by using a USB cable.**

6. **When the tablet is detected on your system, run the application by choosing Run⇨Run or by pressing Ctrl+F11.**

   At this point, the ADT recognizes the device as a new option for a launch configuration (in addition to the emulator); therefore, it displays the Android Device Chooser dialog box to ask you which device it should use to run the application. In Figure 7-9, my Motorola XOOM shows up with a different icon than the emulator to help me identify which is a device and which is an emulator. *Please note:* The emulator won't show up in the list of available options unless the emulator is running.

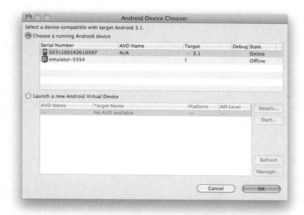

Figure 7-9:
The Android
Device
Chooser.

7. **Choose your device from the list and click the OK button.**

   Doing this sends the application to your tablet, and your app launches just as it would with the emulator. In a few seconds, the app should show up on your device.

Reinstalling Your Application

Installing your application to a physical device is fairly simple. In fact, aside from setting up the device to allow non-Market applications and to allow debugging, installing your app on a physical device requires virtually the same steps as installing it on an emulator.

Well, the same goes for reinstalling your application. When you change or update your app, you need to reinstall it to test it again. Luckily, Android makes doing so pretty easy. You don't have to do anything special to reinstall your application. Just perform the same steps you used to install the application the first time: Choose Run➪Run or press Ctrl+F11. (Easy, huh?)

Uh-oh! — Responding to Errors

You wrote perfect code, right? I thought so! Well, I have a secret to tell: I don't always write perfect code. When things don't go as planned, I have to figure out what is going on. To help developers in the dire situations of random application crashes, the ADT provides some valuable tools to help debug your application, including the Dalvik Debug Monitor Server (DDMS) and the Eclipse debugger.

Using the Dalvik Debug Monitor Server

Debugging is rarely fun. Thankfully, the Dalvik Debug Monitor Server (DDMS) provides the tools necessary to help you dig yourself out of a hole of bugs. DDMS is a debugging tool that provides the following features (among others):

- Port forwarding
- Screen capture
- Thread and heap information on the device
- LogCat (provides dumps of system log messages)
- Process and radio state information
- Incoming call and SMS spoofing
- Location data spoofing

DDMS can work with an emulator and with a connected device. DDMS is located in the Android SDK `tools` directory. In Chapter 4, you added the `tools` directory to your path; therefore, you should be able to access DDMS from the command line.

One of the most commonly used features in DDMS is the LogCat viewer, which lets you view the output of system log messages, as shown in Figure 7-10. This system log reports everything from basic information messages, which include the state of the application and device, to warning and error information. When you receive an Application Not Responding or a Force Close error on the device, what happened is unclear. Opening DDMS and reviewing the entries in LogCat can help identify, down to the line number, where the exception is occurring. DDMS won't solve the problem for you (darn it!), but this tool can make tracking down the root cause of the issue much easier.

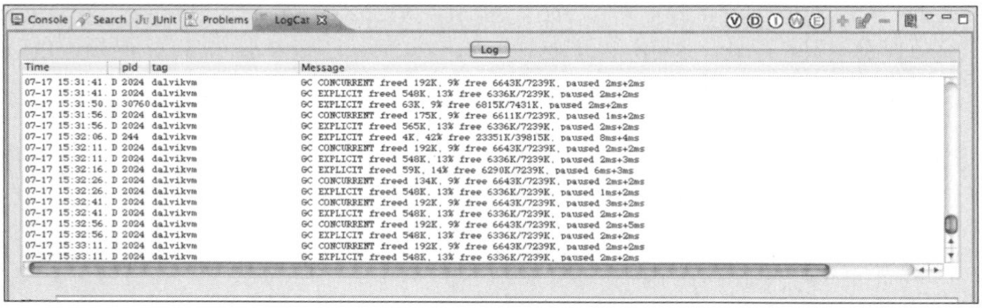

Figure 7-10:
A view of
LogCat.

DDMS is also very useful in scenarios where you do not have an actual device to test with. For example, say you're developing an application that uses GPS and a Google MapView to show the user moving across a map. If you don't have a device that has GPS, or a device at all for that matter, testing this app is a very nontrivial task! Thankfully, DDMS is here to help. DDMS provides location control tools that allow developers to manually supply GPS coordinates or to submit GPS eXchange Format (GPX) or Keyhole Markup Language (KML) files that represent points on a map that can be timed accordingly (for example, stay at this point for 5 seconds, go to this point, and then go to the next point, and so on).

I'm barely scratching the surface of DDMS and its feature set. In the next couple of sections, I show you how to get messages into DDMS as well as how to view them from Eclipse. You can read more about DDMS by going to http://developer.android.com/guide/developing/debugging/ddms.html.

Logging messages into DDMS

Getting messages into DDMS is as simple as supplying one line of code. Open the `MainActivity.java` file, and at the bottom of the method, add a log entry, as shown in Listing 7-7.

Listing 7-7:   The onCreate() Method

```
@Override
public void onCreate(Bundle savedInstanceState) {
    super.onCreate(savedInstanceState);
    setContentView(R.layout.main);

    Log.d("BrightnessToggle", "This is a test");          →6
}
```

The code listed at line 6 demonstrates how to get a message into the system log. SilentModeApp BrightnessToggle is known as the TAG that you're giving this log entry; the second parameter to the log call is the message that you want to output. The TAG helps filter messages while looking at them in DDMS.

A good convention to follow is to declare a TAG constant in your code and use that instead of repeatedly typing the TAG. An example would be

```
private static final String TAG = "BrightnessToggle";
```

Notice the d in Log.d in Listing 7-7. The d means that this is a debug message. Other logging types exist for you to decide how various messages should be logged. Other options are as follows:

- ✔ e: error
- ✔ I: info
- ✔ wtf: What a terrible failure (yes, I'm serious, it's there!)
- ✔ v: verbose

You have to import the android.util.Log package for logging to work.

Viewing DDMS messages

You're probably wondering how you can view the DDMS messages. You can do so in either of these two ways:

- ✔ **Manually, within DDMS:** Navigate to where you installed the Android SDK. Inside the tools directory, double-click the ddms.bat file. Doing this starts the DDMS application outside of the Eclipse IDE, as shown in Figure 7-11.

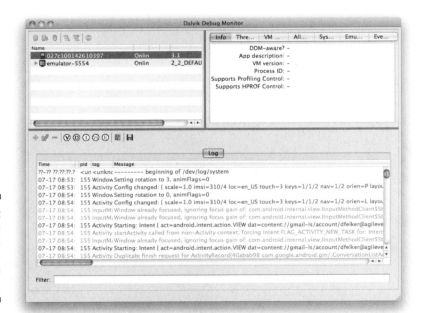

Figure 7-11: An instance of DDMS running separately from Eclipse.

✔ **In Eclipse:** The ADT has installed a DDMS perspective in Eclipse. To open the DDMS perspective, click the Open Perspective button in Eclipse, as shown in Figure 7-12, and choose DDMS. (If DDMS is not visible in this view, select the Other option and then select DDMS. Doing this adds a DDMS perspective to the list of perspectives that you can easily toggle between.) In the DDMS perspective, you can view LogCat (usually near the bottom of the screen). I prefer to move my LogCat window to the main area of the screen, as shown in Figure 7-13. To move your LogCat window to this location, simply drag the LogCat tab title and drop it to the location you want.

Now, start your application by choosing Run⇨Run or by pressing Ctrl+F11. When your application is running in the emulator, open the DDMS perspective and look for your log message. It should look somewhat similar to what is shown in Figure 7-14. The other system log messages may be different on your machine, but the log you typed will be the same as mine.

Open Perspective button

Figure 7-12:
The Open
Perspective
button.

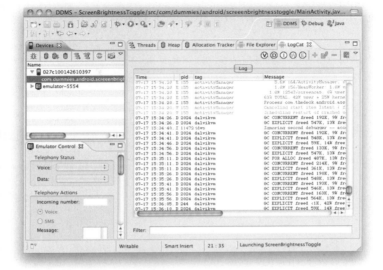

Figure 7-13:
The LogCat window in the main viewing area of Eclipse.

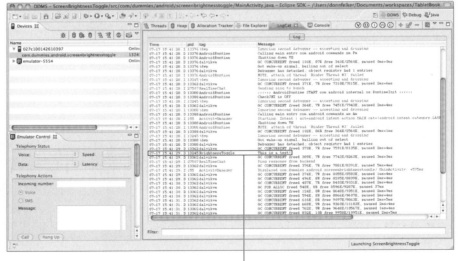

Figure 7-14:
Viewing your LogCat message in Eclipse through the DDMS perspective.

The log message

You can now switch back to the Java perspective by clicking the Java Perspective button, as shown in Figure 7-15.

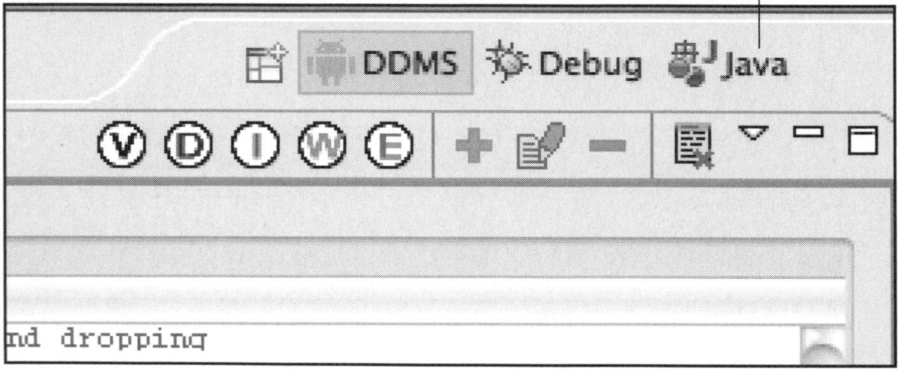

Java Perspective button

Figure 7-15:
Opening the
Java per-
spective.

Using the Eclipse debugger

Although DDMS is one of your best allies, your number-one weapon in the battle against the army of bugs is Eclipse's debugger. This debugger lets you set various breakpoints, inspect variables through the watch window, view LogCat, and much more.

You will use the debugger for either run-time errors or logic errors, but not for syntax errors. Syntax errors are caught by Eclipse during the compiling process. When Eclipse catches a syntax error, your application won't compile, and Eclipse alerts you to the error by placing a colored squiggly line beneath the problematic area.

Checking run-time errors

Run-time errors are like the nasty Wicked Witch of the East. They come out of nowhere and leave everything a mess. In Android, run-time errors occur while the application is running. For example, say your application is humming along just fine, and then all of a sudden, you perform an action, such as clicking a menu or a button, and your application crashes. This is a run-time error. Why did it happen? Perhaps you didn't initialize the ToggleButton in the onCreate() method, and then you tried to access the variable later in the app. This would cause a run-time exception to occur. However, this is only one explanation out of many.

In this situation, the debugger could help determine the problem. If you set a breakpoint at the start of onCreate(), you could inspect the values of the variables through the debug perspective. Tracking the values of these variables could allow you to see what's going on — and to determine that you forgot to initialize the `ToggleButton`.

Listing 7-8 demonstrates the code that would create this scenario. Here, commenting out the `ToggleButton` initialization causes an exception to be thrown at run time.

Listing 7-8:   Commenting Out the `ToggleButton` Initialization

```
ToggleButton tb;                                                          →1
TextView tv;

@Override
    public void onCreate(Bundle savedInstanceState) {
        super.onCreate(savedInstanceState);
        setContentView(R.layout.main);

        //tb = (ToggleButton)findViewById(R.id.toggleButton);              →9
        tv = (TextView)findViewById(R.id.currentBrightness);

        tb.setOnClickListener(new View.OnClickListener() {                →12
            public void onClick(View v) {
                // Code removed for brevity
            }
        });

    // Code removed for brevity
    }
```

These lines do the following:

→**1**   The class-level `ToggleButton` is introduced.

→**9**   Here you "accidentally" comment out this code when doing some testing. This leaves the `tb` variable in a null state.

→**12**   When you attempt to set the click listener for the `ToggleButton`, the application throws a run-time exception. Because `tb` is null, you were trying to reference a member on that object (that did not exist!).

Attaching a debugger to the `onCreate()` method allows you to track down the root cause of the error. The next couple of sections show you how.

In the next few sections, I show you how to set a breakpoint to detect this error. So that your code produces the same error, open your `MainActivity. java` file and comment out line 9 of the `onCreate()` method, as I've done in Listing 7-8.

Creating breakpoints

You have several ways to create a breakpoint:

- ✔ Choose the line where you'd like to insert the breakpoint by clicking it with your mouse. Now, choose Run➪Toggle Breakpoint.

- ✔ Choose the line where you'd like the breakpoint by clicking it with your mouse. Now press Ctrl+Shift+B (for Windows) and Ctrl+Cmd+B (for Mac). This key combination is shown in Figure 7-15.

- ✔ Double-click the left *gutter* of the Eclipse editor where you'd like a breakpoint to be created.

Now use any of these methods to set a breakpoint on line 12 (in Listing 7-8) in your `MainActivity.java` file. This creates a small round icon in the left gutter of the Eclipse editor, as shown in Figure 7-16.

Setting the Debuggable property

You have one last thing to do before you get started with debugging: You have to set up the Android application as debuggable. To do that, open the `AndroidManifest.xml` file, select the Application tab at the bottom, and then locate the Debuggable property (see Figure 7-17) and set it to `true`. Now save the file.

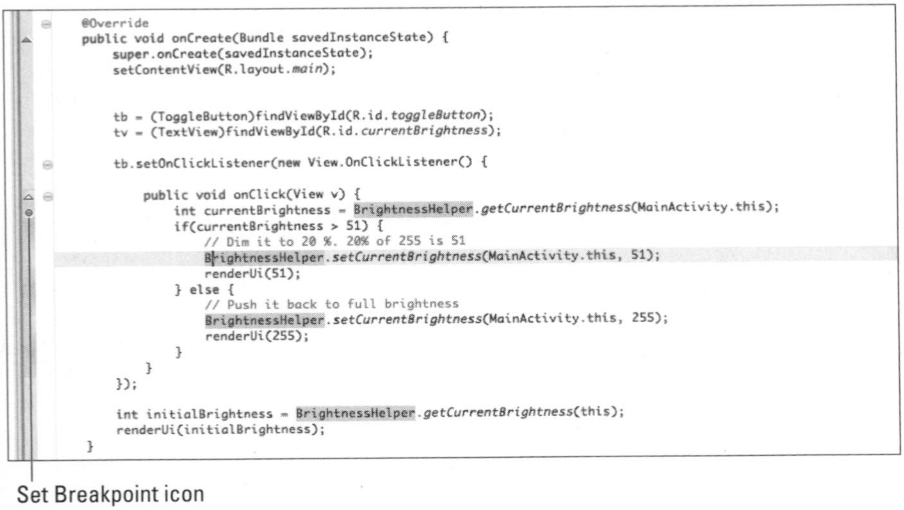

Figure 7-16:
A set breakpoint in the left gutter of Eclipse's editor window.

Set Breakpoint icon

Android Manifest Application

▼ Application Toggle

The application tag describes application-level components contained in the package, as well as general application attributes.

☑ Define an <application> tag in the AndroidManifest.xml

▼ Application Attributes
Defines the attributes specific to the application.

Name		Browse...	Vm safe mode		▾
Theme		Browse...	Hardware accelerated		▾
Label	@string/app_name	Browse...	Manage space activity		Browse...
Icon	@drawable/icon	Browse...	Allow clear user data		▾
Logo		Browse...	Test only		▾
Description		Browse...	Backup agent		Browse...
Permission		▾	Allow backup		▾
Process		Browse...	Kill after restore		▾
Task affinity		Browse...	Restore needs application		▾
Allow task reparenting		▾	Restore any version		▾
Has code		▾	Never encrypt		▾
Persistent		▾	Large heap		▾
Enabled		▾	Cant save state		▾
Debuggable		▾			

true
false

Application Nodes

▶ A .MainActivity (Activity)
A com.dummies.android.screeenbrightnesstoggle.ToggleAct
▶ R com.dummies.android.screeenbrightnesstoggle.AppWidge
S com.dummies.android.screeenbrightnesstoggle.AppWidge

Add...
Remove...
Up
Down

▦ Manifest A Application P Permissions I Instrumentation 🗎 AndroidManifest.xml

Figure 7-17:
Setting the application up as debuggable.

The Debuggable property

Failing to set the Debuggable property to `true` ensures that you never get to debug your application. Your application will not even attempt to connect to the debugger. If I have problems with debugging, I check this place first, because I often forget to set this property to `true`.

Starting the debugger and the Debug perspective

You've created buggy code, you're ready for it to fail, and you're ready to start debugging. I bet you never thought you'd say that out loud!

To start the debugger, choose Run➪Debug or press F11. This tells the ADT and Eclipse to install the application onto the emulator (or device) and then attach the debugger to it.

If your emulator is not already brought to the foreground, do that now. The application installs, and now you see a screen that looks like the one shown in Figure 7-18. This screen informs you that the ADT and the emulator are trying to make a connection behind the scenes.

Figure 7-18:
The emula-
tor waiting
for the
debugger to
attach.

The emulator might sit for a moment while the debugger attaches. After the debugger attaches, it runs your application code and stops when it finds its first breakpoint. Upon doing so, you are presented with a dialog box asking whether it's okay to open the Debug perspective, as shown in Figure 7-19. Click Yes.

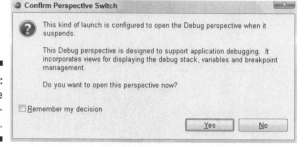

Figure 7-19:
Enabling the
Debug per-
spective.

You should now be at a breakpoint, as shown in Figure 7-20. You can hover over variables to see their values.

Hover your cursor over the `tb` variable, and you find that it is currently null (as shown in Figure 7-20). The reason for this is that commenting out the code, as you did earlier, fails to initialize this variable.

You can also step through the execution of the code by operating the debug navigation, as shown in Figure 7-20. If you click the Resume button (or press F8) three times, you can see the Debug perspective change and eventually say `source not found`. Open the emulator, and you can see that your application has crashed, as shown in Figure 7-21. In the Android Market, users have come to know of this screen as the Force Close, or FC, screen. A force close happens when a run-time exception occurs that is not handled inside your code.

To disconnect the debugger, click the Stop button, as shown in Figure 7-20. Return to the Java perspective, and uncomment line 9 from Listing 7-8 in the `MainActivity.java` file to ensure that the application builds successfully.

Figure 7-20:
The Debug
perspective
explained.

Figure 7-21:
A Force Close dialog box presented due to a run-time exception.

Thinking Beyond Your Application Boundaries

If the device on which your app is running is also downloading a large file in the background while playing music from an online radio application, will these heavy network-bound activities affect the application in any way? Or if your app needs a connection to the Internet and for some reason can't get to the Internet, will it crash? What will happen? Knowing the answers to questions like these is what I refer to as *thinking beyond your application boundaries.*

Not all apps are created equal — and trust me, I've seen some good ones and some *really bad* ones. Before building or releasing your first Android application, you need to make sure that you know the ins and outs of your application and anything that could affect the application. You need to make sure that your app doesn't crash when users perform routine tap events and screen navigation.

Building applications on embedded devices (mobile phones, tablets, and so on) is much different than doing so on a PC or Mac, and the reason is simple: Your resources (memory, processor, and so on) are very limited. To deal with these limitations, Android devices prioritize their duties. If the Android device is a phone, its main purpose is to perform phone-like duties such as recognizing an incoming call, keeping a signal, sending and receiving text messages, and so on. When a phone call is in progress, the Android system treats that process as vital, and other processes — such as downloading a file in the background — as nonvital. If the phone starts to run out of resources, Android will kill off all nonvital processes to keep the vital ones alive.

What if your app contains some of these nonvital processes, and what if one of these processes gets killed? You need to test this scenario. You also need to test for all possible solutions and have a safety guard for them. Otherwise, your app will be prone to run-time exceptions, which can lead to poor reviews on the Android Market.

Ensuring that your app works is as simple as firing up the application and playing with the features. While your app is running, start another app, such as the browser. Surf around the Net for a moment, and then return to your app. Click some button(s) on your app and see what happens. Try all kinds of things to see whether you find outcomes that you may not have thought of. What happens if a user is interacting with your app when an e-mail comes in and they decide to open the e-mail client? Are you saving the necessary state in onPause() and restoring it in onResume()? Android handles the hard task management for you, but it's ultimately your responsibility to manage the state of your application.

Chapter 8

Turning Your Application into an App Widget

*U*sability is the name of the game in application development. When it comes down to it, if your application is not usable, users will not use it. It's that simple.

In Chapter 7, you built the Screen Brightness Toggle application, which works great and is very usable. Unfortunately, if you published the app to the Android Market, the application would not be very popular. Why? In short, because it's not simple enough. Before a user can click the button to change the brightness of the device, he must first go through several awkward steps just to open the app. Unless the user has created a Home screen shortcut to the application, the app is buried in the application launcher with 30 other applications, which means the user must first unlock the device, open the launcher, find the application, and open the app — all before she even gets to the toggle button. This saves the user no time. She might as well just use the settings panel to adjust the screen brightness. So how would you make this application simpler and more usable for the end user? Simple: Turn it into an app widget. That's what I show you how to do in this chapter.

Working with App Widgets in Android

In Android, app widgets (also known as Home screen widgets) are miniature applications that can be embedded within other applications, such as the Home screen. App widgets normally resemble small icons or very small views on your Home screen, and they allow users to interact with your application by simply tapping them.

These app widgets can accept user input through click events and can update themselves on a regular schedule. App widgets are applied to the Home screen by long-pressing (pressing on the screen for a couple of seconds) and then selecting widgets, as shown in Figure 8-1.

To make the Screen Brightness Toggle application more usable, I show you how to build an app widget for the application so that users can add it to their Home screen. After adding the widget, the user can tap it to activate the app's core functionality and toggle the screen brightness mode — without first having to open the application. The widget also updates its layout to inform the user what state the device is in, as shown in Figure 8-2.

Working with remote views

Because Android is based on the Linux 2.6 kernel, the Android platform inherits security idioms from Linux. For example, the Android security model is heavily based around the Linux user, file, and process security model.

Accordingly, each application in Android is (usually) associated with a unique user ID, and all application processes run under a particular user. This prevents one application from modifying the files of another application — which is useful in preventing a malicious developer from injecting code into another app.

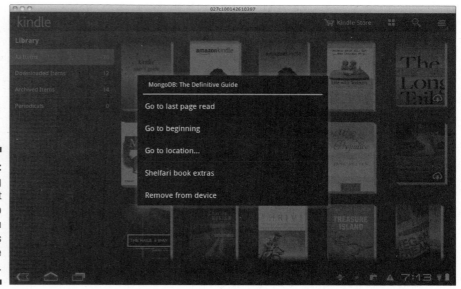

Figure 8-1:
The dialog box that shows up after you long-press the Home screen.

Figure 8-2:
The two
states of the
app widget
you're about
to build.

Evening mode
(20% brightness)

Daytime mode
(100% brightness)

However, this restriction makes it difficult for developers to access or modify the Home screen from within an application. In Android, because the Home screen is actually an application, Android's security restrictions won't allow you as a developer to modify actual running code on the Home screen. Without this rule, malicious developers could do some really evil things, such as shut down your Home screen. How would you use your device then? This is a big issue.

To deal with this problem, the Android engineers implemented the *remote view architecture.* This architecture allows you to run code inside your application, completely away from the Home screen application, while still allowing you to update a view inside the Home screen. The end result is that no arbitrary code can be run inside the Home screen application — all your app widget code runs within your application.

This app widget stuff may sound confusing, but imagine it like this: The user taps the app widget (in this case, an icon on the Home screen that he added). This action fires off a request to change the brightness mode. Android routes that request to your application. During the processing of that request, your application instructs the Android platform to change the brightness mode as well as to update the app widget on the Home screen with a new image to indicate that the brightness mode has been changed. None of this code was run in the Home screen application — all of it was run remotely in your application with Android messaging performing the message routing to the appropriate application.

Remote views (known as the `RemoteView` class in the Android platform) are a little bit of magic mixed with innovative engineering. They allow your application to programmatically supply a remote UI to the Home screen in another process. Structuring things this way means that the app widget code is not an actual activity, as it was in Chapter 7, but is an implementation of an `AppWidgetProvider`. When Android routes a message to your application from the app widget on the Home screen, your implementation of the `AppWidgetProvider` class is where you handle the message.

Using AppWidgetProviders

The `AppWidgetProvider` class provides the hooks that allow you to programmatically interface with the app widget on the Home screen. When the user interacts with the app widget, messages are sent to and from the app widget to your application through *broadcast events.* Through these broadcast events, you can respond to when the app widget is updated, enabled, disabled, and deleted. You can also update the look and feel of the app widget on the Home screen by using a `RemoteView` to update the layout on the Home screen. All the logic that determines what should happen is initiated through an implementation of an `AppWidgetProvider`.

The `AppWidgetProvider` does all the work of responding to events from the `RemoteView`, but how so? `AppWidgetProvider` is a direct subclass of `BroadcastReceiver`, which, at a high level, is a component that can receive broadcast messages from the Android system. When a user taps a clickable view in the `RemoteView` on the Home screen (such as a button), the Android system broadcasts a message informing the receiver that the view was clicked. The message is broadcast to a particular destination in the Android system. After the message is broadcast, the `AppWidgetProvider` can handle that message.

Note that these messages are *broadcast,* meaning that they are sent system-wide. If the message's destination address information is vague enough, a number of `BroadcastReceiver` objects might handle the message, which is not the effect you're looking for. However, the `AppWidgetProvider` you build in this chapter will be addressed to a single destination.

The app widget framework that is built into Android can be thought of as a translator for a conversation between two entities. Imagine that you need to speak to someone who knows Italian, but you don't know Italian. How would you do this? You'd find a translator. The translator would accept your input, translate it to Italian, and relay the message to the native Italian speaker. The same goes for the app widget framework. This framework is your translator.

Here's a great analogy: When the Italian native (the Home screen, in this case) needs to let you know that something has happened (such as a button click), the translator (the app widget framework and the Android system) translates that message into one that you (your app) can understand. At that time, you (your app) can respond with a message instructing the Italian native (the Home screen) to do something (such as changing the app widget background color to lime green), and the translator (the app widget framework) relays the message to the native Italian speaker (through the Android system to the Home screen). When all the translation is done, the Home screen updates the view to have a background color of green.

Because a user's actions are translated through the Android messaging architecture (which requires the apps to communicate across process boundaries), your app will not be immediately notified when the user taps the app widget. However, this does not mean your app won't be notified at all — it just means it's done a little differently.

App widget click events contain instructions on what to do when a click event happens through the use of the PendingIntent class in the Android framework. I explain this in the next section.

Updating the view is about the only thing you can do in regard to app widgets. App widgets can only accept input from tap-type events. You do not have access to other basic input widgets, such as an editable text box, drop-down lists, or any other input mechanism when working within an app widget.

Working with Pending Intents

As I describe in the previous section, the AppWidgetProvider is essentially a translator. It receives a message from the Home screen app widget and responds with a result you specified. The AppWidgetProvider does not work with just any message, though. If you want to receive input from your app widget, you need to use a PendingIntent.

To understand what a PendingIntent is, you need to fully grasp the basic concept of intents and how they work.

Understanding the Android intent system

An Intent object in Android is, well, exactly that: an intent. The best way to think about intents is to envision turning on a light with a light switch. Your intent is to turn on the light, and to do so, you perform the action of flipping the switch to the On position. In Android, this correlates to creating an instance of the Intent class with an Action in it specifying that the light is to be turned on, as shown here:

```
Intent turnLightOn = new Intent("TURN_LIGHT_ON");
```

You fire this intent into the Android messaging system, where an activity (or various different `Activity` objects) responds appropriately. (In the event many activities respond, Android lets the user choose which one to use.) In the case of your `turnLightOn` intent, you could provide code in the form of an activity (named `TurnLightOnActivity`, which is where the code executes) to respond by turning on the light.

As I mention in Chapter 3, an intent is a message that can carry a wide variety of data describing an operation that needs to be performed — that is, an `Activity`. Intents can be addressed to a specific activity or background service or — as I show you in this chapter — broadcast to a generic category of receivers known as `BroadcastReceivers`. When broadcast, the intent will be received only by `BroadcastReceiver` objects that understand how to respond to that intent. In this way, the `Intent`, `Activity`, and `BroadcastReceiver` system is like the well-known messaging structure called *message bus architecture,* in which a message is placed onto a "message bus" and one (or many) of the endpoints on the bus respond to the message if and only if they know how to. If each endpoint has no idea how to respond to the message, or if the message was not addressed to the endpoint, the message is ignored.

An intent can be launched in the following ways:

✔ To start another activity, you use the `startActivity()` call. The `startActivity()` accepts an `Intent` object as a parameter.

✔ To communicate with a background service (covered later in this chapter), you can use the `startService()` or `bindService()` call, which both take intents as parameters.

✔ To notify any interested `BroadcastReceiver` components, you use the `sendBroadcast()` call, which also takes an intent as a parameter.

An activity can be thought of as the glue between various components of the application because it provides a late-binding mechanism that allows inter/intra-application communication.

Understanding intent data

An intent's primary data is as follows:

✔ **Action:** The general action to be performed. A few common actions include `ACTION_VIEW`, `ACTION_EDIT`, and `ACTION_MAIN`. You can also provide your own custom action if you choose to do so.

✔ **Data:** The data to operate on, such as a record in a database or a uniform resource identifier that should be opened, such as a website URL.

Table 8-1 demonstrates a few action and data parameters for `Intent` objects and their simple data structure.

Table 8-1	Intent Data Examples	
Action	*Data*	*Result*
ACTION_EDIT	content://con-tacts/people/1	Edits the information about the person whose given identifier is 1
ACTION_VIEW	http://www.exam-ple.org	Displays the web page of the given intent
ACTION_VIEW	content://con-tacts/people	Displays a list of all the people in the contacts system

For a list of some of the predefined intents in the Android system, see `http://developer.android.com/guide/topics/intents/intents-filters.html`.

Intents can also carry an array of other data that include the following:

- ✔ **category:** Gives additional information about the action to execute. As an example, if `CATEGORY_LAUNCHER` is present, it means that the application should show up in the application launcher as a top-level application. Another option is `CATEGORY_ALTERNATIVE`, which can provide alternative actions that the user can perform on a piece of data.

- ✔ **type:** Specifies a particular type (MIME type) of the intent data. Setting the type to `audio/mpeg`, for example, would indicate to the Android system that you are working with an MP3 file. Normally the type is inferred by the data itself. By explicitly making this setting in the intent, you override the inferred type inference.

- ✔ **component:** Specifies an explicit component name of the class to execute the intent upon. Normally, the component is inferred by inspection of other information in the intent (action, data/type, and categories), and the matching component(s) can handle it. If this attribute is set, none of that evaluation takes place, and this component is used exactly as specified. You'll probably want to specify a component in most of your applications. Providing another activity as the component tells Android to interact with that specific class.

- ✔ **extras:** A bundle of additional information that is key based. This bundle is used to provide extra information to the receiving component. For example, if you'd like your application to have the ability to interact with the built-in e-mail client, you would need to set up an intent that would provide the e-mail address and would use the extras bundle to supply the body, subject, and other components of the e-mail. After you started the intent, the Android intent system would recognize these attributes via the intent filter system and would start the e-mail client on the device. For more information on intent filters, see `http://developer.android.com/guide/topics/intents/intents-filters.html`.

Intents are evaluated in the Android system in one of two ways:

- ✔ **Explicitly:** The intent has specified an explicit component or the exact class that will execute the data in the intent (again, this will probably be the most common way you address intents). These types of intents often contain no other data because they are a means to start other activities within an application. I show you how to use an explicit intent in this application later in the chapter.

- ✔ **Implicitly:** The intent has not specified a component or class. Instead, the intent must provide enough information about the action that needs to be performed with the given data for the Android system to determine which available components can handle the intent. This portion of the intent is sometimes referred to as an *address* and *payload*.

  An example of this is setting up an e-mail intent that contains e-mail fields (To, CC, Subject, and Body) and an e-mail MIME type. Android interprets the intent as an e-mail intent and gives the user of the device the opportunity to choose which application should handle the intent — Gmail, Exchange, or a POP e-mail account that are all enabled on the device. This allows the user to determine from where the e-mail should originate. Android's feature of identifying the possible matches for the given intent is known as *intent resolution*.

Using pending intents

On one level, a `PendingIntent` acts just like a regular intent. However, a `PendingIntent` allows the intent to be performed by another application.

A `PendingIntent` is created by your application and given to another completely different application. By giving another application a `PendingIntent`, you grant the other application the right to perform the specified operation as if the application was your own. In other words, a `PendingIntent` allows your application to perform work on another application's behalf. When an application executes the `PendingIntent`, it instructs the Android messaging system to inform your application to perform the necessary work.

For our purposes, to obtain a pending intent instance, I use the `PendingIntent.getBroadcast()` call. This call returns a `PendingIntent` that is used for broadcasts throughout the system. The call takes four parameters:

✔ `Context`: The context in which this `PendingIntent` should perform the broadcast.

✔ `RequestCode`: The private request code for the sender. Not currently used; therefore, a zero is passed in.

✔ `Intent`: The intent to be broadcast.

✔ `Flags`: A set of controls used to control the intent when it is started. Not currently used; therefore, a zero is passed in.

Wait a second — this looks a bit funky. This code uses an `Intent` as well as a `PendingIntent`. Why? The `Intent` object is wrapped inside a `PendingIntent` because a `PendingIntent` is used for cross-process communication. When the `PendingIntent` is fired off, the real work that needs to be done is wrapped up in the child `Intent` object.

When a user taps the button on the Home screen, the Home screen "app" sends a `PendingIntent` to the `AppWidgetProvider` in your application. At that time your app responds according to the instructions contained in the `Intent` that is part of the `PendingIntent` (Brighten screen, dim, and so on). The app then brightens the screen and sends instructions to the app widget via the `AppWidgetManager` to update its icon. The `AppWidgetManager` updates the icon via a `RemoteView`, which also contains a new `PendingIntent` (which contains a regular `Intent`) that will be executed the next time the user taps the Home screen widget.

Whoa, that was a lot of information! Now that you understand the basics of the Android intent system, you can implement the guts of the application inside this app widget.

Creating the App Widget

A lot is going on when it comes to interacting with an app widget. The process of sending messages between the home app widget and your application is handled through the Android messaging system, the `PendingIntent` class, and the `AppWidgetProvider`. In this section, I demonstrate how to build each component so that you can get your first app widget up and running on your Home screen.

Implementing the AppWidgetProvider

Implementing the `AppWidgetProvider` is a straightforward process. Just follow these steps:

1. **Open Eclipse and open the Screen Brightness Toggle application.**

2. **Add a new class to the `com.dummies.android.screenbrightness toggle` package and provide a name; I prefer to use `AppWidget.java`.**

   To add a new class, right-click `com.dummies.android.screen-brightnesstoggle` in the `src/` folder and choose New⇨Class. This opens the New Java Class dialog box. In this dialog box, provide the name of the class and set its superclass to `android.appwidget.AppWidgetProvider`, as shown in Figure 8-3.

3. **Click Finish.**

   A new class has been added to the selected package, and the code with the name you chose should now be visible.

Figure 8-3:
The New
Java Class
dialog box.

Communicating with the app widget

Right now, your `AppWidgetProvider` class has no code in it — it's an empty shell. For your `AppWidgetProvider` to do anything, you need to add a bit of code to respond to the intent (the message) that was sent to your `AppWidgetProvider`. In the code file you just created, type the code shown in Listing 8-1 into the editor. (***Note:*** My class is called `AppWidget.java`, so if yours is different, you need to change that line.)

Listing 8-1:   The Initial Setup of the App Widget

```
public class AppWidget extends AppWidgetProvider {                    →1
    @Override
    public void onReceive(Context ctxt, Intent intent) {             →3
            if (intent.getAction()==null) {                         →4
             // Do Something
            } else {                                                →6
                super.onReceive(ctxt, intent);                      →7
            }
        }
}
```

Here is a brief explanation of what the various lines do:

→**1**   This line of code informs Android that this class is an `AppWidgetProvider` because the class is inheriting from `AppWidgetProvider`.

→**3**   This line overrides the `onReceive()` method to be able to detect when a new intent is received from the `RemoteView`. This intent could have been initiated by a user tapping a view to perform an action such as a button click. The `Intent` object is contained within the `PendingIntent`, which initiated the request.

→**4**   As described previously, `Intent` objects can contain various pieces of data. One such slice of data is the action. On this line of code, I am checking to see whether the intent has an action. If it does not have an action, I know that I fired off the intent. This may sound a bit backward, but I explain why this is done in the upcoming sections.

→**6**   An action was found in the `Intent` object; therefore, a different logical sequence of events needs to take place.

→**7**   You delegate the work to the superclass because you don't need to do anything with the intent because it isn't what you were expecting (the intent had an action — you were expecting an intent without an action). This would happen if the app widget automatically updated itself on a regular period that you defined in widget metadata (explained in the "Working with the app widget's metadata" section later in this chapter).

Building the app widget's layout

The app widget needs to tell Android how to display the widget on the Home screen, and it does so through the widget layout file. Earlier in the chapter, I show the real app widget's two different icons. Although you can't see it in Figure 8-2, those icons are surrounded by a background — a transparent one, which blends perfectly with the natural Home screen background. This layout was defined by the widget layout file. If you were to change the background color of the layout file from transparent to lime green, the background color of the widget on the Home screen would also change to lime green, as shown in Figure 8-4.

Figure 8-4: The result of changing the widget's background color to lime green.

This lime green background also illustrates the concept of *widget screen space*. The lime green box shown in Figure 8-4 identifies the available screen space for the app widget. You can define your app widget to take up one Home screen cell or many cells. (This app widget only takes up one cell.)

To create your widget layout, create an XML layout file in the `res/layouts` directory and add the contents of Listing 8-2. When you're finished, you can name your file whatever you like — I'm going to name mine `widget.xml`.

Listing 8-2:   The widget layout file — `widget.xml`

```
<?xml version="1.0" encoding="utf-8"?>
<RelativeLayout xmlns:android="http://schemas.android.com/apk/res/android"
    android:layout_width="fill_parent"
    android:layout_height="fill_parent">
    <ImageView android:id="@+id/brightnessState"
        android:layout_height="wrap_content"
        android:layout_width="wrap_content"
        android:layout_centerInParent="true"
        android:src="@drawable/icon"
        android:clickable="true" />                              →10
</RelativeLayout>
```

This layout is nothing that you have not seen before. It's a `RelativeLayout` that has one child view: a clickable `ImageView`. You can click this `ImageView` by setting the clickable property to `true` on line 10 in Listing 8-2. (For more on `RelativeLayout`, see Chapter 6.)

Note the `src` property of the `ImageView`. It is set to the icon of the application. I'm sure that seems a bit odd to you, but here is why I did that. When I built the layout, I had not yet created the brightness state buttons that represented bright and dim states. However, I did need to preview what the view would look like in the layout designer while designing the layout. Therefore, I used `@drawable/icon` as the value of the `ImageView` to glean some vision of how the view is going to look. The fact that I'm using the application icon does not concern me at this point because when the app widget loads, the `ToggleService` switches the icon value to either the bright or dim mode state icon, as shown later in this chapter.

These icons help the end user of the application identify the current state of the application. The `sun` icon signifies when the device is in a daytime brightness mode. The `night` icon signifies when the device is in the evening brightness mode. I created these icons in an image-editing program. You can create your own, or you can use mine by downloading them from this book's companion website.

Adding Images to Your Application

Up to this point you've been able to work without adding any new images to your application. Now you need to put some stuff on the screen! Although looking at text is fun and all, the real interesting components are added through input mechanisms and images. In the following sections, I demonstrate how to include images in your application.

Placing an image on the screen

First, add the brightness mode state images to the application so you can use them later. These are the widget image state icons shown in Figure 8-2, earlier in this chapter. First you need (uh, yeah) the images. You can create your own images, or you can download mine from this book's companion website.

Adding images to your project is simple. You simply drag them over and then reference them in the project.

Drag the images into the `res/drawable-hdpi` folder in the Eclipse project, as shown in Figure 8-5.

Figure 8-5:
Dragging
the image
file into
the res/
drawable-
hdpi folder.

Notice that you see two states of the widget: daytime and evening, as shown earlier in Figure 8-2.

When you dragged the images into Eclipse, the ADT recognized that the project file structure changed. At that time, the ADT rebuilt the project because the Build Automatically selection is enabled in the Project menu. This regenerated the `gen` folder, where the `R.java` file resides. The `R.java` file now includes a reference to the two new images that were recently added. You may now use these references to these resources to add images to your layout in code or in the XML definition. You declare them in the XML layout in the following section.

Adding the image to the layout

You've already added an image to the layout, but in that instance you were using the `icon` for the application. What if you wanted to use a different image? To do that, you would need to type some code into your layout file (an example of this is shown here). *Note:* Do not add this code to your widget layout file — it is for demonstration purposes only.

```
<?xml version="1.0" encoding="utf-8"?>
<LinearLayout xmlns:android="http://schemas.android.com/apk/res/android"
    android:orientation="vertical"
    android:layout_width="fill_parent"
    android:layout_height="fill_parent"
    >
    <ImageView
            android:id="@+id/my_image"                              →8
            android:layout_width="wrap_content"
            android:layout_height="wrap_content"
            android:layout_gravity="center_horizontal"              →11
            android:src="@drawable/sun" />                          →12

</LinearLayout>
```

Why you should worry about density folders

Android supports devices with various screen sizes and *densities,* which are like screen resolutions. Just now, you placed an image in the hdpi folder, which is for high-density devices. What about small- and medium-density devices? If Android cannot find the requested resource in the desired density, it opts for a density of the resource that it can find. What does this mean? If you're running on a high-density screen, a lower-density image will be chosen, and the image will appear stretched out and pixilated. If you're running on a low-density device, a higher-density image will be chosen, and the image will be compressed to fit within the screen dimensions. To avoid this, create multiple versions of your image to target each screen density. For more information, see the Supporting Multiple Screens best practice guide in the Android documentation located at `http://developer.android.com/guide/practices/screens_support.html`.

In this step, you added the `ImageView` inside the `LinearLayout`. An `ImageView` allows you to project an image to the device's screen.

The `ImageView` contains a couple of extra parameters that you have not seen yet, so I cover those now:

→**8** **The `android:id="@+id/my_image"` property:** The id attribute defines the unique identifier for the view in the Android system. For an explanation of the `android:id` value nomenclature, try the actual Android documentation on the subject, which is located at `http://developer.android.com/guide/topics/ui/declaring-layout.html#id`.

→**11** **The `layout_gravity` property:** This property defines how to place the view within its parent. Here, I have defined the value as the `center_horizontal` constant. This value informs the Android system to place the object in the horizontal center of its container, and not to change its size. You can use many other constants, such as `center_vertical`, `top`, `bottom`, `left`, `right`, and many more. See the `LinearLayout.LayoutParams` Android documentation for a full list. More about LayoutParams can be found at `http://d.android.com/reference/android/widget/LinearLayout.LayoutParams.html`.

→**12** **The `android:src="@drawable/sun"` property:** This property is a direct child of the `ImageView` class. You use this property to set the image that you would like to show up on the screen.

Notice the value of the `src` property — `"@drawable/sun"`. What you're seeing now is the use of the `R.java` file. Here, you can reference `drawable` resources through XML. This is done by typing the at symbol (@) and the resource you're after.

And when you type in that resource, you don't have to bother specifying its density. (In Listing 8-2, I did not type `@drawable-mdpi` for the drawable resource identifier — I typed `@drawable`.) This is because it is Android's job (not yours) to support multiple screen sizes (which makes life easy for you!). The Android layout system knows about drawables — and that's all. It knows nothing of low-, medium-, or high-density drawables during design time. At run time, Android determines whether it can use low/medium/high-density drawables and when it should use them.

For example, if the app is running on a high-density device and the requested drawable resource is available in the `drawable-hdpi` folder, Android uses that resource. Otherwise, it uses the closest match it can find. Supporting various screen sizes and densities is a large topic (and complex in some aspects). Therefore, for an in-depth view into this subject, read the Managing Multiple Screen Sizes best practice article in the Android documentation.

The `sun` portion of line 12 identifies the drawable that you want to use. The image filename is actually `sun.png`. However to stay within Java's member-naming guidelines, you must remove the file extension when you specify this file, leaving `sun`. If you were to open the `R.java` file in the `gen` folder, you would see a member variable with the name of `sun`, not `sun.png`.

Thanks to the ADT, you can see your available options for this property through code completion. Place your cursor directly after `"@drawable/"` in the `src` property of the `ImageView` in the Eclipse editor. Then press Ctrl+spacebar. You should see the code completion dialog box, as shown in Figure 8-6. See the other resource names in there? These are other options that you could also choose for the `src` portion of the drawable definition.

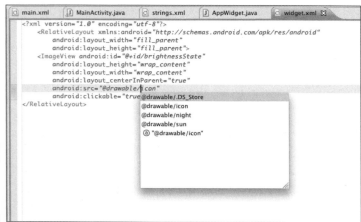

Figure 8-6:
Code completion with resources.

Creating a Launcher Icon for the Application

When your app gets installed, an icon for it appears in the application launcher to help users identify its presence. When you created the Screen Brightness Toggle application, the ADT automatically included a default launcher icon for you, as shown in Figure 8-7.

Figure 8-7: The default Android launcher icon.

Kind of bland, right? Change this icon to one of your own. I have a round sunshine icon that I created in an image-editing program, as shown in Figure 8-8. You can create your own (as shown in the following section) or download mine from this book's companion website.

Figure 8-8: The new sunshine application icon.

Designing a custom launcher icon

Creating your own launcher icons is fairly easy thanks to the Android project. The Android documentation contains a full article, entitled "Icon Design Guidelines, Android 2.0," that includes a how-to manual for creating icons for the Android platform, along with a style guide, dos and don'ts, materials and colors, size and positioning guidelines, and best of all, icon templates that you can use. You can find all of the design guidelines at `http://d. android.com/guide/practices/ui_guidelines/icon_design.html`.

Working with templates

Because you've already downloaded the Android SDK, icon templates and materials are available for you to use right now on your hard drive! Navigate to your Android SDK installation directory (from Chapter 2), and from there navigate to the `docs/shareables` directory. There you'll find various `.zip` files that contain templates and samples. Open the templates in the image-editing program of your choice and follow the design guidelines in the documentation to make your next rocking icon set.

If you want to really speed up things, you can use the Android Asset Studio, located here: `http://android-ui-utils.googlecode.com/hg/asset-studio/dist/index.html`. This tool helps you quickly create great-looking icons for the application, menu, or notification bar as well as many other graphic assets for your application.

Matching icon sizes with screen densities

You think about multiple densities when working with images, and icons are no different. You need to create three different versions of your icon (low-density, medium-density, and high-density) for the icon to show up correctly on various devices. Because each screen density requires a different-size icon, you must size your icon correctly, so it will look appropriate (no pixelation, stretching, or compressing) on the screen. Table 8-2 summarizes the finished icon sizes for each of the three generalized screen densities.

Table 8-2	Finished Icon Sizes
Screen Density	*Finished Icon Size*
Low-density screen (ldpi)	36 x 36 px
Medium-density screen (mdpi)	48 x 48 px
High-density screen (hdpi)	72 x 72 px

Adding a custom launcher icon

To place your custom launcher icon into the project, follow these steps:

1. **Rename your custom icon's filename to** `icon.png`.

2. **Drag the three correctly sized versions of your icon into the** `mdpi`, `hdpi`, **and** `ldpi` **folders.**

   Eclipse asks whether you want to overwrite the existing `icon.png` file, as shown in Figure 8-9. Click Yes.

Figure 8-9:
Eclipse asks
whether
it's okay to
overwrite
the icon.
png file.

If you don't copy the low- and high-density icons into their respective folders, users who have a low- or high-density device receive the default launcher icon, as shown in Figure 8-7, earlier in this chapter, whereas the medium-density devices receive the new icon that you dragged into the project.

How does this happen? You dragged the file into the mdpi folder; what gives? The hdpi and ldpi folders both contain their own version of the icon. Open the drawable-hdpi and drawable-ldpi folders in your Eclipse project and you can see that each density has its own icon.png file. Be sure to place the correct-size icon in each density-specific folder.

Handling Application Logic inside an AppWidgetProvider

After the pending intent has started your AppWidgetProvider, you need to perform some logic on behalf of the calling application (in this case, the Home screen application). In the following sections, I show you how to perform time-sensitive work on behalf of the caller.

Before I hop into the code, it's best to understand how the work will be done in the AppWidgetProvider. Due to the nature of remote processes and how resource-intensive they can be, it's best to do all work inside a background service. I will be performing the changing of the brightness mode via a background service.

Understanding the IntentService

Why should you use a background service for such a trivial task as changing the device brightness mode? Well, it has to do with avoiding Application Not Responding (ANR) errors.

Any code that executes for too long without responding to the Android system is subject to the ANR error. App widgets are especially vulnerable to ANR errors because they are executing code in a remote process — they execute across process boundaries that can take time to set up, execute, and tear down — the entire process is very CPU-, memory-, and battery-intensive. The Android system watches app widgets to ensure that they do not take too long to execute. If they do take an extended period of time to execute, the calling application (the Home screen) locks up and the device is unusable. Therefore, the Android platform wants to make sure that you're never capable of making the device unresponsive for more than a couple of seconds.

Because app widgets are very expensive in regard to CPU and memory, it's really hard to judge whether an app widget will cause an ANR error. If the device is not doing any other expensive tasks, the app widget would probably work just fine. However, if the device is in the middle of one or many expensive CPU operations, the app widget could take too long to respond — causing an ANR error. This unknown CPU state is a dangerous variation to introduce to your app widget. Therefore, to get around it, move the work of the app widget into a space where it can take as long as it needs to complete, a space that in turn will not affect the Home screen application — in other words, a background service. (You're using a background service called an `IntentService` for this app.) Unlike most background services, which are long-running, an `IntentService` uses the work queue processor pattern that handles each intent in turn using a worker thread, and it stops when it runs out of work. In laymen's terms, the `IntentService` simply takes the work given to it, runs it as a background service, and then stops the background service when no more work needs to be done.

Implementing the AppWidgetProvider and IntentService

In your `AppWidgetProvider` class, type the code in Listing 8-3 into your code editor.

Listing 8-3:   The Full `AppWidget` Implementation

```
public class AppWidget extends AppWidgetProvider {

    @Override
    public void onReceive(Context context, Intent intent) {

        if (intent.getAction()==null) {
            context.startService(new Intent(context, ToggleService.class));
                                                                          →7
            Intent i = new Intent(context, ToggleActivity.class);         →8
            i.addFlags(Intent.FLAG_ACTIVITY_CLEAR_TASK);                  →9
            i.addFlags(Intent.FLAG_ACTIVITY_NEW_TASK);                    →10
```

```
        context.startActivity(i);                                    →12
    } else {
        super.onReceive(context, intent);                            →14
    }
}

public static class ToggleService extends IntentService {            →18

    public ToggleService() {
        super("AppWidget$ToggleService");                            →21
    }

    @Override
    protected void onHandleIntent(Intent intent) {                   →25
        ComponentName me=new ComponentName(this, AppWidget.class);   →26
        AppWidgetManager mgr=AppWidgetManager.getInstance(this);     →27
        mgr.updateAppWidget(me, buildUpdate(this));                  →28


    }

    private RemoteViews buildUpdate(Context context) {              →32
        RemoteViews updateViews=new
            RemoteViews(context.getPackageName(),R.layout.widget);   →34

        int currentBrightness =
            BrightnessHelper.getCurrentBrightness(ToggleService.this); →37
        if(currentBrightness > 51) {                                 →38
            // Dim it to 20 %. 20% of 255 is 51
          updateViews.setImageViewResource(R.id.brightnessState,
             R.drawable.night);                                      →41
        } else {
            // Push it back to full brightness
               updateViews.setImageViewResource(R.id.brightnessState,
               R.drawable.sun);                                      →45
        }

        Intent i=new Intent(this, AppWidget.class);                  →48
        PendingIntent pi = PendingIntent.getBroadcast(context, 0, i,0);
                                                                     →49
        updateViews.setOnClickPendingIntent(R.id.brightnessState,pi); →50

        return updateViews;
    }
  }
}
```

The following list briefly explains what each major section of code does:

→**7**   This line of code starts a new instance of the ToggleService.
           The context object in this line of code refers to the Android
           Context object, which is an interface to global information

about the application. The context is passed into the onReceive() and onUpdate() method calls. A new intent is created to let the Android system know what should happen. This method is initiated by the user when the user taps the app widget on the Home screen.

→8   This line builds a new intent that you will use to start an activity.

→9   This line sets one of the two flags that will be set on this intent. The first flag is the FLAG_ACTIVITY_CLEAR_TASK. This flag informs the Android system to clear any existing task associated with the activity prior to starting a new one.

→10   This line sets the second of the two flags that will be set on this intent. The second flag is the FLAG_ACTIVITY_NEW_TASK. This flag informs the Android system to start a new task on the Activity history stack.

→12   This line starts the activity. You may be wondering why I'm starting an activity from the widget intent service. Remember back when you built the MainActivity class (in Chapter 7) when you changed the current window brightness by setting the layout parameters with the brightness level? Without that call, the brightness setting in the system would change, but the current screen brightness would not. Therefore, you have to start the ToggleActivity for about one second to update the brightness of the device while the activity is running. This is needed because this is the only way that you can get a handle on the actual running window of the device. This activity has not been created yet. I will cover this new activity below.

→14   Because the intent action was not null, a default course of action was taken to allow the superclass to handle the onReceive event.

→18   This is an implementation of an IntentService. This IntentService handles the same logic as your MainActivity for handling the brightness-mode switching but in regard to the app widget infrastructure. This is an implementation of a background service in the Android platform, as described previously. This class is a nested static class within the app widget.

→21   This method calls the superclass with the name AppWidget$ ToggleService. This method call takes place to help with debugging purposes for the thread name. If you leave this line of code out, you receive a compiler error informing you that you must explicitly invoke the super's constructor. If your app widget is named something else other than AppWidget, you should change this to the class name of your class.

→25   The onHandleIntent() method is responsible for handling the intent that was passed to the service. In this case, it would be the intent that was created on line 7. Because the intent that you created was an explicit intent (you specified a class name to execute), no extra data was provided, and therefore by the time you get to line 24, you don't need to use the intent anymore. However, you could have provided extra information to the `Intent` object that could have been extracted from this method parameter. In this case, the `Intent` object was merely a courier to instruct the `ToggleService` to start its processing.

→**26** A `ComponentName` object is created. This object is used with the `AppWidgetManager` (explained next) as the provider of the new content that will be sent to the app widget via the `RemoteViews` instance.

→**27** An instance of `AppWidgetManager` is obtained from the static `AppWidgetManager.getInstance()` call. The `AppWidgetManager` class is responsible for updating the state of the app widget and provides other information about the installed app widget. You will be using it to update the app widget state.

→**28** The app widget gets updated with a call to `updateAppWidget()` on this line. This call needs two things: the Android `Component Name` that is doing the update and the `RemoteView` object that is used to update the app widget. The `ComponentName` is created on line 24. The `RemoteView` object that will be used to update the state of the app widget on the Home screen is a little more complicated and is explained next.

→**32** The method definition for the `buildUpdate()` method. This method returns a new `RemoteView` object that will be used on line 28. The logic for what should happen and the actions to proceed with are included in this method.

→**34** Here I am building a `RemoteView` object with the current package name as well as the layout that will be returned from this method. The layout, `R.layout.widget`, is shown in Listing 8-2.

→**37** The code from line 37 to line 45 is very similar to the code that you wrote for the `MainActivity`. The code between these lines checks the current brightness level of the device and then decides which code path to execute. This line gets the current brightness level of the device using the `BrightnessHelper` you built previously.

→**38** If the brightness level is greater than 51, then the device is assumed to be in bright mode.

→**41** This line of code updates the `RemoteView`'s `ImageView` whose ID is `brightnessState`. The new image will be the evening image (`night.png`).

→**45** If the logic determines that the brightness level is not greater than 51, then this code will execute. This code updates the `RemoteView`'s `ImageView` whose ID is `brightnessState`. The new image will be the daytime image (`sun.png`).

→**48**   This line creates an `Intent` object that will start the `AppWidget` class when initiated.

→**49**   Unfortunately, app widgets cannot communicate through basic intents; they require the use of a `PendingIntent`. Remember, app widgets use cross-process communication; therefore, `PendingIntent` objects are needed for communication. On this line, I build the `PendingIntent` that instructs the app widget of its next action via the child intent built on line 48.

→**50**   Because you're working with a `RemoteView`, you have to rebuild the entire event hierarchy in the view. The reason for this is the app widget framework will be replacing the entire `RemoteView` with a brand-new one that you supply via this method. Therefore, you have one thing left to do: tell the `RemoteView` what to do when it's tapped/clicked from the Home screen. The `PendingIntent` that you built on line 49 instructs the app widget what to do when someone clicks or taps the view. The `setOn-ClickPendingIntent()` sets this up. This method accepts two parameters: the ID of the view that was clicked (in this case an image) and the `pi` argument, which is the `PendingIntent` that you created on line 49. In other words, you're setting the click listener for the `ImageView` in the app widget.

Adding the ToggleActivity to Help Adjust Current Window Brightness

In Listing 8-3 you were introduced to the `ToggleActivity`. The `Toggle Activity` has one function — to act as an invisible courier that does work, kind of like a ninja on a midnight mission! The `ToggleActivity` checks the current screen brightness and then, depending upon the current state of that brightness, updates the current window parameters with the new screen brightness in order to get the brightness to change. You want this activity to start, do its work so fast that no one notices, and then close. You do not want the user to view the `Activity` — they should not notice a new activity starting. Therefore, you will hide the title bar of the activity. When you register this activity later in the chapter in the `AndroidManifest.xml` file, you will inform the Android system to make the background of the activity transparent, effectively hiding any known trace that the activity is actually running.

Why does this step need to be done in an activity? Why can't it be done in a service? Because you need access to the currently running window in order to get access to the current window/screen brightness. A background service runs in the background, so it has no window. Because the widget is running in a remote process, Linux security won't let you have access to that window. Therefore, you have to start a new activity (one hidden from the user), so you can gain access to the window in order to change the current brightness.

To set up your activity, add a new class with the name of `ToggleActivity.java` to the `src/` folder. Replace the contents of this file with what is shown in Listing 8-4.

Listing 8-4:   The ToggleActivity.java

```
public class ToggleActivity extends Activity {

    @Override
    public void onCreate(Bundle savedInstanceState) {
        super.onCreate(savedInstanceState);
        requestWindowFeature(Window.FEATURE_NO_TITLE);              →6

        int currentBrightness =
                    BrightnessHelper.getCurrentBrightness(this);    →9

        if(currentBrightness > 51) {                                →11
            // Dim it to 20 %. 20% of 255 is 51
            BrightnessHelper.setCurrentBrightness(this, 51);
        } else {
            // Push it back to full brightness
            BrightnessHelper.setCurrentBrightness(this, 255);
        }

        // Wait for 1000 milliseconds (1 second) for the brightness
        // change to start, then quit.
        final Timer t = new Timer();                                →21
        t.schedule(new TimerTask() {                                →22

            @Override
            public void run() {                                     →25
                finish();                                           →26
                t.cancel();                                         →27
            }
        }, 1000);                                                   →29
    }
}
```

In the following list, I explain each major area of the code:

→**6**　　　This line informs the Android system to not show the title bar at the top of this Activity when it starts.

→**9**　　　Gets the current screen brightness so that your code can decide what action to pursue.

→**11**　　This line checks the current brightness and lets execution fall into setting the brightness to evening (dim) if it is greater than 51. Otherwise the code block will fall through to update the brightness to daytime (full brightness) code block.

→**21**　　This line creates a new timer for you to use in the application. You need to use a timer because in order for the screen brightness update to be applied, your application — not `ToggleActivity` — must be in the forefront of the activity stack. If it is not, the screen brightness change will not happen. By creating a timer to schedule when your activity should finish, you ensure that `ToggleActivity` quits in a timely manner. It should only take about 100–300 milliseconds to update the screen brightness, but just to be safe, I've set the value to 1000 (on line 29) of the time schedule call that takes place on line 22. This increase gives the timer 1000 milliseconds (1 full second) to update the screen brightness. After 1000 milliseconds, the activity will be finished (the `ToggleActivity` will quit and will return to the Home screen where the widget is located).

→22   This line calls the schedule method of the `Timer` class. This method allows you to schedule code to be run after a specified time has elapsed. After the time has elapsed, the run method will be called. Any code you put in the run method will be called at that time.

→25   This line defines the run method for the timer's schedule method call.

→26   At this point, the timer's 1000 milliseconds would have elapsed and now your code would be running again. This line calls the activity's `finish` method. The `finish` method finishes (quits) the activity.

→27   Because you are no longer in need of the timer (you just quit the activity by calling its `finish` method), you can now cancel the timer so that it no longer fires.

→29   This line contains the time in which you'd like the specified timer to execute. In this instance, it's going to run every 1000 milliseconds (every second).

You've written the code to implement the widget as well as to update the screen brightness in the settings as well as on the screen. Now you need to work with the metadata of the widget and add the new components to the `AndroidManifest.xml`.

Working with the app widget's metadata

Now that the code is written to handle the updating of the app widget, you may be wondering how to get the app widget to show up on the Widgets menu so that you can install it. This fairly simple process requires that you add a single XML file to your project. This XML file describes some basic metadata about the app widget so that the Android platform can determine how to lay out the app widget onto the Home screen. Here's how you do this:

1. **In your project, right-click the** res **directory and choose New⇨New Folder.**

2. **For the folder name, type** xml **and click Finish.**

3. **Right-click the new** res/xml **folder, choose New, and then choose Android XML File.**

4. **In the New Android XML File Wizard, type** widget_provider.xml **for the filename.**

5. **The file type will be of the type** AppWidgetProvider. **Select that radio button and then click Finish.**

6. **After the file opens, open the XML editor and type the following into the** widget_provider.xml **file:**

```xml
<?xml version="1.0" encoding="utf-8"?>
<appwidget-provider xmlns:android="http://schemas.android.com/apk/res/
        android"
        android:minWidth="79px"
        android:minHeight="79px"
        android:updatePeriodMillis="0"
        android:initialLayout="@layout/widget"
/>
```

The minWidth and minHeight properties are used for setting the very minimum space that the view will take on the Home screen. These values could be larger if you want.

The updatePeriodMillis property defines how often the app widget should attempt to update itself. In the case of the Screen Brightness Toggle application, you do not want this to happen — the only time the brightness should change is if the user initiates it. Therefore, this value is set to 0 seconds. If you were to set this to 180000 milliseconds — 30 minutes then every 30 minutes, the app would update itself through sending an intent that executes the onUpdate() method call in the AppWidgetProvider (if you have implemented that method).

The initialLayout property identifies what the app widget will look like when the app widget is first added to the Home screen before any work takes place. Note that it may take a couple of seconds (or longer) for the app widget to initialize and update your app widget's RemoteView object by calling the onReceive() method.

An example of a longer delay is if you had an app widget that checked Twitter for status updates. If the network is slow, the initialLayout would be shown until updates were received from Twitter. Therefore, if you foresee this becoming an issue, inform the user in the initialLayout that information is loading. Therefore, the user is kept aware of what is happening when the app widget is initially loaded to the Home screen. You can do this by providing a TextView with the contents of "Loading . . ." while the AppWidgetProvider does its work.

At this point, you can install the Screen Brightness Toggle application, long-press the Home screen, and choose the Widgets category; now you should see the Screen Brightness Toggle present. The metadata that you just defined is what made this happen. The icon defaults to the application icon. However, the app widget would throw an exception if you attempted to add it to the Home screen. This mistake is fairly common: I forgot to let the `AndroidManifest.xml` file know about my new `IntentService` and `BroadcastReceiver`. If the `AndroidManifest.xml` does not know about these new items, exceptions will be thrown because the application context has no idea where to find them. You solve this problem next.

Registering your new components with the manifest

Anytime you add an `Activity`, `Service`, or `BroadcastReceiver` (as well as other items) to your application, you need to register them with the application manifest file. The application manifest presents vital information to the Android platform, namely, the components of the application. The `Activity`, `Service`, and `BroadcastReceiver` objects that are not registered in the application manifest will not be recognized by the system and will not be able to be run. Therefore, if you added the app widget to your Home screen, you would have it crash because your `AppWidgetProvider` is a `BroadcastReceiver`, and the code in the receiver is using a service that is also not registered in the manifest.

To add your `AppWidgetProvider` and `IntentService` to your application manifest file, open the `AndroidManifest.xml` file and type the code shown in Listing 8-5 into the already-existing file. Bolded lines are the newly added lines for the new components.

Listing 8-5:   An Updated `AndroidManifest.xml` File with New Components Registered

```
<?xml version="1.0" encoding="utf-8"?>
<manifest xmlns:android="http://schemas.android.com/apk/res/android"
    package="com.dummies.android.screeenbrightnesstoggle"
    android:versionCode="1" android:versionName="1.0">
    <uses-permission android:name="android.permission.WRITE_SETTINGS"></uses-
            permission>
    <application android:icon="@drawable/icon" android:label="@string/app_name">
        <activity android:name=".MainActivity" android:label="@string/app_name">
            <intent-filter>
                <action android:name="android.intent.action.MAIN" />
                <category android:name="android.intent.category.LAUNCHER" />
            </intent-filter>
        </activity>
```

```
        <activity android:name=
            "com.dummies.android.screeenbrightnesstoggle.ToggleActivity"    →14
            android:theme="@android:style/Theme.Translucent" />             →15
        <receiver android:name=
            "com.dummies.android.screeenbrightnesstoggle.AppWidget"         →17
            android:label="@string/app_name" android:icon="@drawable/icon"
            android:previewImage="@drawable/night">                         →19

    <intent-filter>
        <action android:name="android.appwidget.action.APPWIDGET_UPDATE" /> →22
    </intent-filter>

            <meta-data android:name="android.appwidget.provider"
                android:resource="@xml/widget_provider" />                  →26
        </receiver>
        <service        android:name=                                      →28
        "com.dummies.android.screeenbrightnesstoggle.AppWidget$ToggleService" />

    </application>
    <uses-sdk android:minSdkVersion="11" />
</manifest>
```

The following is a brief explanation of what each section does:

→**14**   This line registers the `ToggleActivity` so that your application
          can call it.

→**15**   This line instructs the Android system to use a different theme
          for this activity only. The theme that this activity uses is the
          Translucent theme. Therefore, the user will not even see an activ-
          ity start.

→**17**   This line is the opening element that registers a `Broadcast`
          `Receiver` as part of this application. The `name` property
          identifies what the name of the receiver is. In this case, it is
          `.AppWidget`, which correlates to the `AppWidget.java` file in the
          application. The name and the label are there to help identify the
          receiver.

→**19**   This line defines the `previewImage` attribute for the widget.
          When the widget is added to the screen, this is going to be the
          preview image that is shown while the application is loading. A lot
          of popular apps have a preview image that shows a loading icon
          to illustrate that the application widget is indeed loading (perhaps
          the widget must perform some action when it's first added to the
          screen, such as determine the user's location through GPS).

→**22**   Identifies what kind of intent (based on the action of the intent
          in the intent filter) the app widget will automatically respond
          to when the particular intent is broadcast. This is known as an

IntentFilter, and it helps the Android system understand what kind of events your app should get notified of. In this case, your application is concerned about the `APPWIDGET_UPDATE` action of the broadcast intent. This event fires after the `update- Period Millis` property has elapsed, which is defined in the `widget_ provider.xml` file. Other actions include `enabled`, `deleted`, `disabled`, and more. Currently you have this set to zero, instructing the app widget framework to *not* send an update to the provider. However, if you omit this intent filter, you will not be able to add the widget to the home screen.

→**26**   This line identifies the location of the metadata that you recently built into your application. Android uses the metadata to help determine defaults and lay out parameters for your app widget.

→**28**   This line identifies the Intent Service that you created to handle offloading of tasks from the main user interface thread.

At this point, your application is ready to be installed and tested. To install the application, choose Run⇨Run or press Ctrl+F11. Your application should now show up on the emulator. Return to the Home screen by pressing the Home key. You can now add the app widget you recently created to your Home screen.

Placing Your Widget on the Home screen

The usability experts on the Android team did a great job by allowing application widgets to be easily added to the Home screen. Adding a widget to the Home screen is super easy. Follow these steps:

1. **Long-press the Home screen on the emulator by clicking the left mouse button on the Home screen and keeping the mouse button pressed.**

2. **When the Add to Home Screen dialog box is visible, select Widgets, as shown in Figure 8-10.**

3. **When the Choose Widget dialog box is visible, choose Screen Brightness Toggle, as shown in Figure 8-11.**

Figure 8-10: The Add to Home Screen view.

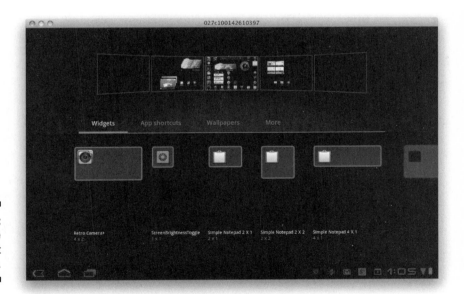

Figure 8-11: The Choose Widget view.

You have now added the Screen Brightness Toggle widget to your Home screen, as shown in Figure 8-12. You can tap the icon to change the brightness mode, and the background will change accordingly, as shown in Figure 8-2.

Figure 8-12:
The app widget added to the Home screen.

Chapter 9

Understanding Android Resources

. . . .

In This Chapter

▶ Knowing why resources are so important in Android

▶ Extracting resources

▶ Working with image resources

. . . .

*T*he information about resources and their usage covered in Chapters 5 and 8 were necessary to help you understand the basics of the resource directory as well as to see how resources are used to build a simple application. However, to succeed as an app developer, you need to understand resources in more detail, so in this chapter, I provide thorough coverage of the topic. You have many compelling reasons to use resources in your application, one being globalization — which I cover in this chapter.

Understanding Resources

Resources are no fly-by-night type of Android idiom. They're first-class citizens in the Android platform. In Android, resources can be any of the following:

✔ Layouts	✔ Themes
✔ Strings	✔ Values
✔ Images	✔ Menus
✔ Dimensions	✔ Colors
✔ Styles	

You've already been introduced to layouts, strings, and images because they are the most common types of resources that you will use in everyday Android application development. The other resources may be a little muddy for you, so let me clear those up here.

Dimensions

In an Android resource, a *dimension* is a number followed by a unit of measurement, such as 10px, 2in, or 5sp. You use dimensions when specifying any property in Android that requires a numeric unit of measure (as when you, say, specify the padding of a layout to be 10px). The following units of measure are supported by Android:

- **dp (density-independent pixels):** I use this unit of measure most when I'm developing my layouts. This abstract unit is based on the physical density of the screen. These units are relative to a 160-dots-per-inch (dpi) screen; therefore, 1 dp is equivalent to one pixel on a 160-dpi screen. The ratio of dp to pixels changes with screen density, but not necessarily within proportion. The dp topic is quite in-depth and should be investigated if you plan to actively support multiple screen densities. *Density* is similar to screen resolution except that density refers to the number of pixels you can squeeze onto the screen. Squeezing more pixels into the screen space (essentially by making pixels smaller) changes a high-resolution screen into a high-density screen. You can read more information about this topic at Supporting Multiple Screen Sizes, located here: `http://developer.android.com/guide/practices/screens_support.html`.

- **sp (scale-independent pixels):** This unit is like the dp unit but is scaled according to the user's font-size preference. You should use sp dimensions when specifying font sizes in your application.

- **pt (points):** A point is ½ inch, based on the physical size of the screen.

- **px (pixels):** These correspond to actual pixels on the screen. This unit of measure is not recommended because your app may look great on a medium-density device but look very distorted and out of place on a high-density screen (and vice versa) because the dpi differs on both devices.

- **mm (millimeters):** Based on the size of the screen.

- **in (inches):** Based on the physical size of the screen.

Styles

Styles allow you to — well, you guessed it — *style* your application! In Android, styles are very similar to Cascading Style Sheets (CSS) in the web development realm. A *style* is a collection of properties that can be applied to any individual view (within the layout file) or activity, or to your entire application (from within the manifest file). Styles support inheritance, so you can provide a very basic style and then modify it for each particular use you have in your application. Example style properties include font size, font color, and screen background.

Themes

A *theme* is a style applied to an entire activity or application, rather than just an individual view. When a style is applied as a theme, every view in the activity or application inherits the style settings. For example, setting all `TextView` views in the theme to be a particular font makes all views in the themed activity or application display text in that font.

Values

Values can contain many different types of value type resources for your application, including the following:

- **Bool:** A Boolean value defined in XML whose value is stored with an arbitrary filename in the `res/values/<filename>.xml` file, where *<filename>* is the name of the file. An example is `bools.xml`.

- **Integer:** An integer value defined in XML whose value is stored with an arbitrary filename in the `res/values/<filename>.xml` file. An example is `integers.xml`.

- **Integer array:** An array of integers defined in XML whose set of values is stored with an arbitrary name in the `res/values/<filename>.xml` file, where *<filename>* is the name of the file. An example is `integers.xml`. You can reference and use these integers in your code to help define loops, lengths, and so on.

- **Typed array:** A typed array is used to create an array of resources, such as `drawables`. You can create arrays of mixed types, so the arrays are not required to be homogeneous — however, you must be aware of the array item data type so that you can appropriately cast it. As with the others, the filename is arbitrary in the `res/values/<filename>.xml` file. An example is `types.xml`.

Menus

Menus can either be defined through code or through XML. The preferred way to define menus is through XML, so the various menus you create should be placed into the `menus/` directory. Each menu has its own `.xml` file.

Colors

The `values/colors.xml` file allows you to give colors a name, such as `login_screen_font_color`, to, say, define the color of the font you use in the logon page. Each color is defined as a hexadecimal value.

Working with Resources

You've worked with resources a few times in this book already, and it's probably familiar to you at this point to use the R class to access resources from within your application. If you're still a bit rusty on resources and the generated R file, see Chapter 5.

As you work more and more with resources, you find a number of common problems. For one thing, it's pretty easy to overlook the tedious chore of inserting strings into your resources. For another, it can be difficult to size images correctly. The next two sections show you how to deal with each of these problems. And the last section shows you how to do something with resources that's very useful — how to make your apps understood around the globe.

Moving strings into resources

When developing a project, I've been known to take a few shortcuts. Because of this, at times I've forgotten to put strings into resources, and I've had to come back later to do this. Because this is a pretty common error, I've actually done this on purpose with the Screen Brightness Toggle application. Here I walk you through the process of inserting a string into a resource using the built-in tools.

Before I do that, though, let me show you the "long way":

1. **Create a new string resource.**

2. **Copy its name.**

3. **Replace the string value in your layout with the resource identifier.**

Okay, this may not be a huge pain, but it takes time, possibly 30–45 seconds for the average developer. Now I show you how to cut that number to under 15 seconds. If you do this 30 times a day (which is feasible in an 8-hour day), you can save 15 minutes of copying and pasting. That's a savings of five hours a month! Follow these steps:

1. **If Eclipse is not open, open it now and open the** `main.xml` **file in the** `layouts` **directory.**

2. **Find the following chunk of code in the file:**

```
<ToggleButton
        android:id="@+id/toggleButton"
        android:layout_width="wrap_content"
        android:layout_height="wrap_content"
        android:layout_gravity="center_horizontal"
        android:textOn="Dimmer On"
        android:textOff="Dimmer Off"
/>
```

3. **Select the boldface line** `"Dimmer On"`.

4. **Press Shift+Alt+A.**

   This step opens a menu with three options.

5. **Choose the Extract Android String option.**

   Doing this opens the Extract Android String dialog box, as shown in Figure 9-1, which allows you to set various options for the resource.

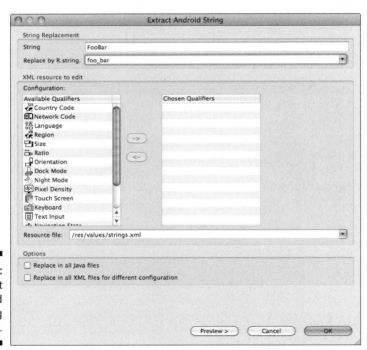

Figure 9-1:
The Extract Android String dialog box.

I'm not going to use any of those features for this, so leave the defaults as they are and click the OK button.

You can now see that the layout file has been modified. The text `"Dimmer On"` has been replaced with `"@string/dimmer_on"`. If you open the `strings.xml` file in the `res/values` folder, you can see a new string resource with that name, and the value of `"Dimmer On"`.

Now, that's pretty cool! You can see that doing this 20–30 times a day can add up and save you a lot of time.

Working with images

Dealing with images can be one of the most difficult parts about resources. Images that look great on a medium-density device may look like garbage on a high-density device. This is where the multiple-density folders come into play. These density-specific drawable folders are explained in Chapter 8.

To get around the issue of pixilation and compression/expansion (when going from higher- to lower-density devices and vice versa), design your graphics at a very high resolution, such as 300 dpi in large-size format. Downsizing a high-density image does not distort the quality (other than losing the fine edges and detail), but upscaling does — because it creates pixilation and distortion. Starting with a large file reduces the chances that you'll ever have to upscale, which means that your app graphics will always look crisp.

For example, if you're building your launcher icon, build it at 250px height and 250px width. Although the `hdpi` folder might need only a 72px-height-x-72px-width image (which is the largest used right now), new Android devices are coming out all the time. In two to three months Google TV may come out, and on a big TV screen, your 72x72 images will look pretty awful.

I know that working with large image files in image-editing programs can be difficult if you don't have a decent-performing computer, but you have to trust me: A large raw image file that is high density is much easier to mold and shape later into the correct densities that you'll need.

Also, if you're creating your graphics in an image-editing tool that supports layers, I highly advise you to place each item in your graphic on a different layer. The reasons for this are many, but here are the key factors:

✔ **Changes:** At some time, you will need to change something in your graphic — maybe the background, maybe the font, maybe the logo. If you have all these items in different layers, you can do that without affecting the rest of the graphic.

> ✔ **Localization:** You will often encounter graphics with stylized text in the graphic itself. If your application is being translated into Japanese, and your graphics contain stylized English text, you'll want to create a Japanese version of those graphics and place them in a Japanese drawable region folder such as `res/drawable-ja`. The Android platform recognizes which region it is in (in this case, Japan). If the region's resource folders (`res/drawable-ja`, `res/values-ja`, and so on) are available, Android uses those in the application.

Making your apps global with resources

In 2010, the Android platform surpassed Apple's iPhone in U.S. market share, trailing only Research In Motion's BlackBerry, according to ZDNet. Now carriers around the world are developing Android-based smartphones, which means one simple thing: more potential users for your apps.

So what does this mean to you as a developer? It means that Android is a huge market with tons of opportunity waiting to be tapped. This opportunity is very exciting, but to take the greatest advantage of it, you need to understand resources and how they affect the usability of your apps. For example, if your app was written for an English audience (using resources or not), a user in the United States would be able to use it. However, if you hardcoded all of your English string values into your views and activities and then decided to release a Chinese version, you would have to rewrite your application. When you use resources, translations such as this are easy.

Resources allow you to extract human-readable strings, images, and viewable layouts into resources that you can reference. Various resource folders can be created to handle various-size screens, different languages (think strings and drawables), and layout options, such as landscape or portrait. Landscape and portrait views come into play when a user rotates the device 90 degrees in either direction.

If you want your apps to be viewable on as many Android devices as possible around the world, you want to use resources at all times. As an example, I advise that you always put all your strings into the `strings.xml` file because someday, someone from another country will want your application in another language. To get your application into another language, you simply need to have a translator translate the text in your `strings.xml` file into the new language, and then you can create various `values` folders to hold the appropriate region's values. For example, if the user is using a phone set to the Chinese character set, Android knows to look for a values folder in your app called `values-cn`, which is where Chinese values are stored — including the Chinese version of the `strings.xml` file. If Android cannot find such a folder, the platform defaults to the default `values` folder, which contains the English version of the `strings.xml` file. (For more on strings, see the section "Moving strings into resources," earlier in this chapter.)

When it comes down to it, having a translator update your strings and adding the new `strings.xml` file to a new values folder are very simple things to do (compared with recoding your entire application). Take this process and expand it to other languages and devices and eventually Google TV . . . and you can see where I'm going. You're no longer looking at mobile users as your target audience. You're looking at Android users, and with the options coming out — this could be billions of users. Using resources correctly can make your expansion into foreign markets that much easier.

Designing your application for various regions is a big topic. You can find more in-depth information in the Localization article of the SDK documentation here: `http://d.android.com/guide/topics/resources/localization.html`.

Although designing your application to be ready for various regions sounds compelling, it also helps to know that the Android Market allows you to specify a targeted region for your app. You're not forced into releasing your application to all regions. Therefore, if you have written an application for the Berlin bus route system in Germany, it probably doesn't make sense to have a Chinese version, unless you want to cater to Chinese tourists as well as German residents. I cover the Android Market in depth in Chapter 10.

Chapter 10

Publishing Your App to the Android Market

The Android Market is the official application distribution mechanism behind Android. Publishing your application to the market enables your application to be downloaded, installed, and enjoyed by millions of users across the world. Users can also rate and leave comments on your application, which helps you identify possible use trends as well as any problematic areas users might be encountering.

You had a great idea and you developed the next best application/game for the Android platform; now you're ready to get the application into the hands of end users. First, you need to package your application so that it can be placed on end users' devices. Android packages are known as *APK files;* in this chapter, I guide you through the creation of your first APK file.

After you obtain an APK file, I walk you through the process of creating an Android Market account and publishing a free application. I also show you how you can sell your application on the Android Market, so you can finally make that first million you've been telling everyone about!

Creating a Distributable APK File

Before you jump in and create the distributable APK, you should take great care to make sure that your application is available to as many users as possible. This is done with the `uses-sdk` element in the `AndroidManifest.xml`. Your `AndroidManifest.xml` file currently has a `uses-sdk` entry that you created in Chapter 8, which is shown here:

```
<uses-sdk android:minSdkVersion="4" />
```

The `minSdkVersion` property in this line identifies which versions of the Android platform can install this application. In this instance, level 4 has been selected. However, for your Screen Brightness Toggle application, you set the target SDK to version 11. So version 4 is the minimum SDK even though you're targeting the version 11 SDK? How can all of this madness work?

The Android platform, for the most part, is backwards compatible. Most of the features in version 11 are also in version 12. Yes, small changes — and sometimes even large components — exist between these versions, but almost everything else in the platform remains backwards compatible. Therefore, stating that this application needs a minimum of SDK version 11 only signifies that any Android operating system that is of version 11 or greater can run the application.

Using the `minSdkVersion` information, the Android Market is able to determine which applications to show each user of each device. If you were to release the application right now with `minSdkVersion` set to the value of 4, and you opened the Android Market on an Android device running version 3 (Android 1.5) or lower, you would not be able to find your application. Because you specified version 4, the Android Market filtered your app out for earlier devices. If you were to open the Android Market on a device running version 4 or above, however, you would be able to find and install your application.

If you do not provide a `minSdkVersion` value in the `uses-sdk` element of the Android Manifest, the Android Market will not allow you to upload your application. Please be sure to provide a valid `minSdkVersion` prior to uploading your app to the Android Market.

You can build an Android APK file in numerous ways:

- ✔ Through the Android Development Tools (ADT) inside of Eclipse
- ✔ By using an automated build process through a continuous integration server, such as Hudson
- ✔ Through the command line with Ant
- ✔ . . . and other ways such as the Maven build system, and so on

The last three of these options are used in advanced scenarios. For the purposes of this chapter, you use the ADT within Eclipse to create your APK file. The ADT provides an array of tools that compiles and packages your Android application files into an APK. As it does this, the ADT "digitally signs" your app. (Digital signing is discussed in the next section.)

Digitally signing your application

The Android system requires that all installed applications be "digitally signed" with a certification that contains a public/private key pair. The private key is held by the developer. The certification is used to identify the application and the developer and for establishing the trust relationships between applications.

There are a few key important things to know about signing Android applications:

- ✔ **All Android applications *must be signed*.** The system will not install an application that is unsigned.
- ✔ **You can use self-signed certificates to sign your applications.** A certificate authority is not needed.
- ✔ **When you are ready to release your application to the market, you must sign it with a private key.** You cannot publish the application with the debug key that creates the APK when debugging the application during development.
- ✔ **The certificate has an expiration date that is verified only during install time.** Therefore, if the certificate expires after the application has been installed, the application will continue to operate normally.
- ✔ **You don't have to use the ADT tools to generate the certificate.** You can use standard tools, such as `Keytool` or `Jarsigner` to generate and sign your APK files.

Applications signed with the same certificate can communicate with each other. This allows the applications to run within the same process, and, if requested, the system can treat these applications as a single application. With this methodology, you can create your application in modules, and users can update each module as they see fit. A great example of this would be to create a game and then to release "update packs" to upgrade the game. Users can decide to purchase only the updates they like.

The certificate process is outlined in detail in the Android documentation. The documentation includes how to generate certificates with various tools and techniques.

Creating a keystore

In Android (as well as in Java), a *keystore* is a container that holds your personal certificates. You can create a keystore file with a couple of different tools:

✔ **The ADT Export wizard:** This tool is installed with the ADT and allows you to export a self-signed APK file that will digitally sign the application as well as create the certificate and keystore (if needed) through a wizardlike process.

✔ **The keytool application:** This application allows you to create a self-signed keystore through the command line. This tool is located in the Android SDK `tools/` directory and provides many options through the command line.

In this chapter, you use the ADT Export wizard to create your keystore.

The keystore file contains your private certificate, which Android uses to identify your application in the Android marketplace. If you lose your keystore, you will not be able to sign the application with the same private key, which can cause problems. For one thing, this prevents you from upgrading your app — the Android Market will recognize your file as a new Android application, not as an upgrade to an existing one, and won't allow you to proceed. This also happens if you change the package name of the app — Android will not recognize your file as a valid update. To make sure this doesn't happen, back up your keystore in a safe location.

Creating the APK file

To create your APK, follow these steps:

1. **Open Eclipse (if it is not already open).**

2. **Right-click on Screen Brightness Toggle, choose Android Tools, and then choose Export Signed Application.**

   The Project Checks window of the Export Android Application wizard appears, as shown in Figure 10-1.

3. **In the Project Checks window, click Next.**

   Doing this opens the Keystore Selection window, as shown in Figure 10-2.

4. **Because you have not yet created a keystore, select the second radio option, Create New Keystore. In the Location field, type the location of your keystore — I prefer to use `c:\android\` (on Windows) or `~/android` (for Mac OS X and Linux) — and a name for your keystore. The filename should have the `.keystore` extension.**

   My full path looks like this:

   ```
   c:\android\dummies.keystore
   ```

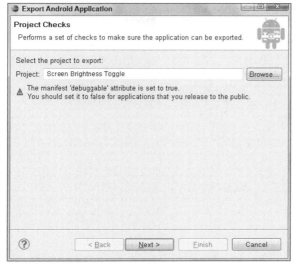

Figure 10-1:
The Project
Checks
window of
the Export
Android
Application
wizard.

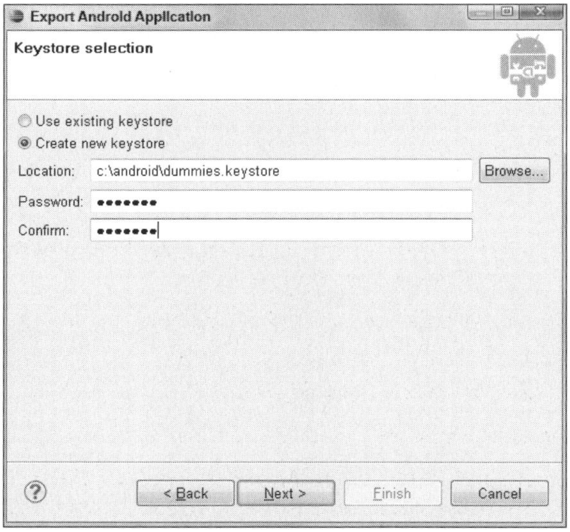

Figure 10-2:
The
Keystore
Selection
window.

5. Choose a password that you can remember and type it in the Password and the Confirm fields, and then click Next.

Doing this opens the Key Creation page, which allows you to create the keystore.

6. **In the Key Creation page, fill out the following fields and then click Next.**

   • **Alias:** This is the alias that you will use in the future to identify the key.

   • **Password & Confirmation:** The password that will be used for the key. You can use the same password as in the Keystore selection or you can use a new one.

   • **Validity:** The length of time this key will be valid. It is required that your key expire after 22 October of 2033. I normally plug a value of 30 years into this field to be safe.

7. **Complete the certificate issuer section of the dialog box by filling out at least one of these fields and then click Next:**

   • First and Last Name

   • Organization Unit

   • Organization

   • City or Locality

   • State or Province

   • Country Code (XX)

   I have chosen to provide my name as the issuer field. When you finish, your dialog box should resemble Figure 10-3.

Figure 10-3:
The Key
Creation
window.

The final screen of the Export Android Application wizard, the Destination and Key/Certificate Checks window, appears (see Figure 10-4).

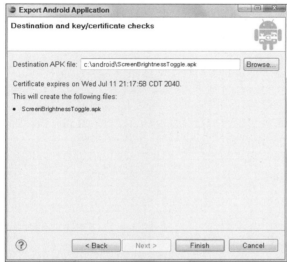

Figure 10-4: Choosing a filename and destination for your first APK.

8. **In the Destination and Key/Certificate Checks window, you specify a destination for your APK file. Please choose a name and location for your file (with an extension of .apk), and type it in the Destination APK File field.**

   The name and location I chose is

   ```
   c:\android\ScreenBrightnessToggle.apk
   ```

9. **Click Finish.**

   Doing this creates the .apk file in your chosen location as well as a keystore in the location you chose in Step 4. (See Figure 10-5.)

You have now created a distributable APK file and a reusable keystore for future updates.

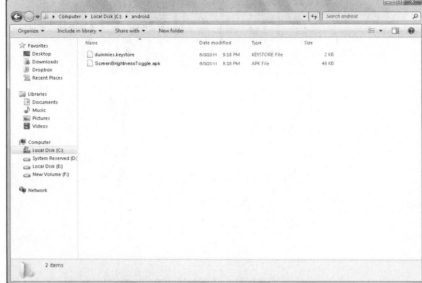

Figure 10-5:
The new
APK and
keystore in
the same
destination
folder.

Creating an Android Marketplace Account

Now that you have your APK file created, you can release your application on the Android Market. However, before you do so, you first need to create an Android Market account. To create such an account you need a Google account — any Google-based account (such as a Gmail account) will do. If you do not have a Google account, you can obtain a free one by navigating your web browser to `www.google.com/accounts/` and following the steps there.

Please note that to create an Android Market account, you must pay a $25 developer fee with a valid credit card. If you do not pay this developer fee, you will not be able to publish applications.

After you have a Google account, use the following steps to create your Android Market account:

1. **Open your web browser and navigate to** `http://market.android.com/publish`.

2. **On the right side of the screen, sign in with your Google account, as shown in Figure 10-6.**

   When you're logged in, you can complete the fields to define your developer profile.

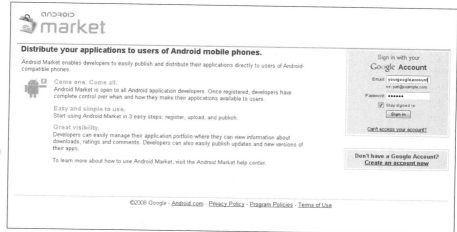

3. **Complete the following fields and then click the Continue link (see Figure 10-7):**

   • **Developer Name:** The name that shows up as the developer of the applications you release. The developer name can be your company name or your own personal name. You can change this later, after you've created your account.

   • **Website URL:** The URL of your website. If you do not have a website you can get a free Blogger.com blog that will suffice as a website. You can get a free Blogger account from `www.blogger.com`.

   • **Phone Number:** A valid phone number to contact you at in case users encounter problems with your published content.

Figure 10-7:
Developer
listing
details.

4. **On the next page, click the Continue link to go to Google Checkout to pay the developer fee. (See Figure 10-8.)**

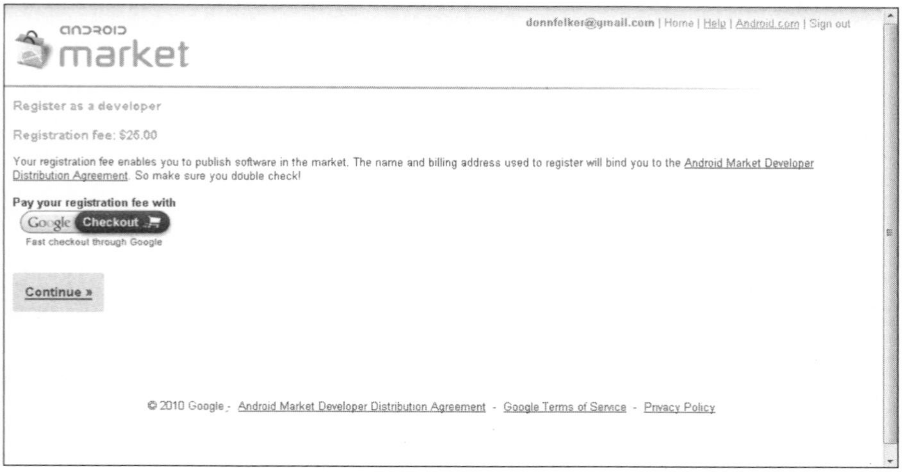

Figure 10-8:
Developer
registration
fee.

5. **On the secure checkout page, fill in your credit card details and billing information and click Agree and Continue. (See Figure 10-9.)**

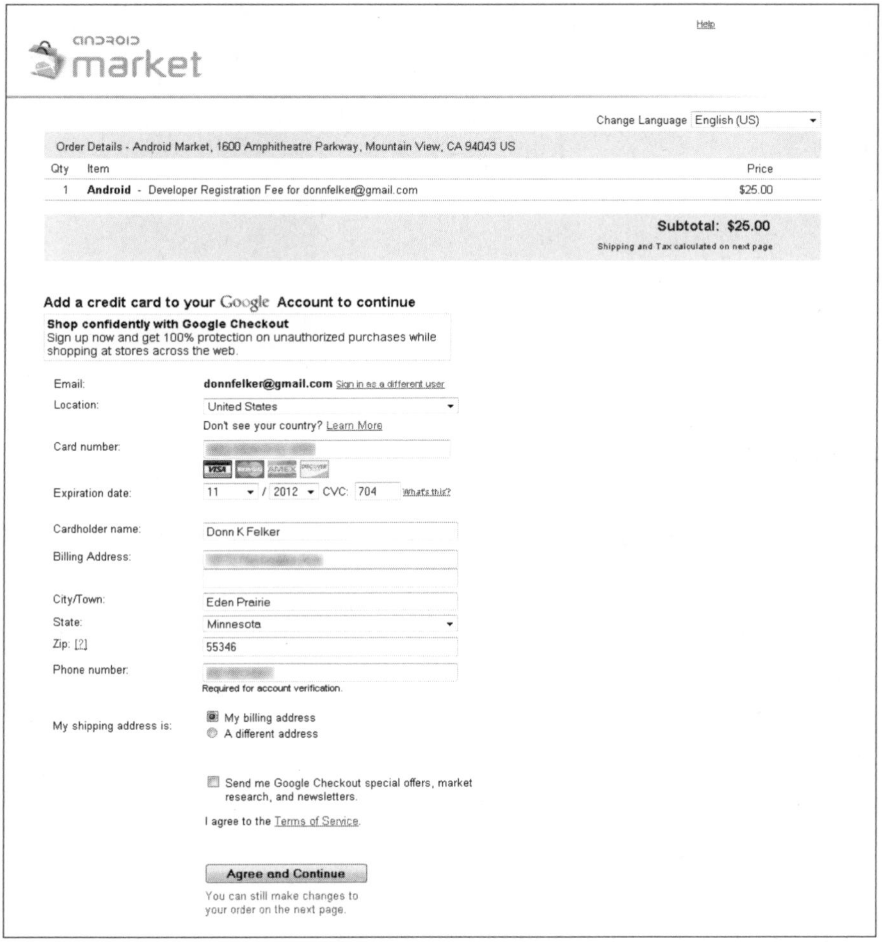

Figure 10-9:
Personal
and billing
information.

6. **To confirm your choice, type in your password and click Sign In and Continue. (See Figure 10-10.)**

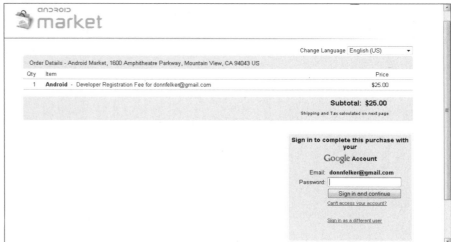

Figure 10-10:
The sign-in
page for
registering
as a
developer.

7. **On the order confirmation page (see Figure 10-11), click Place Your Order Now.**

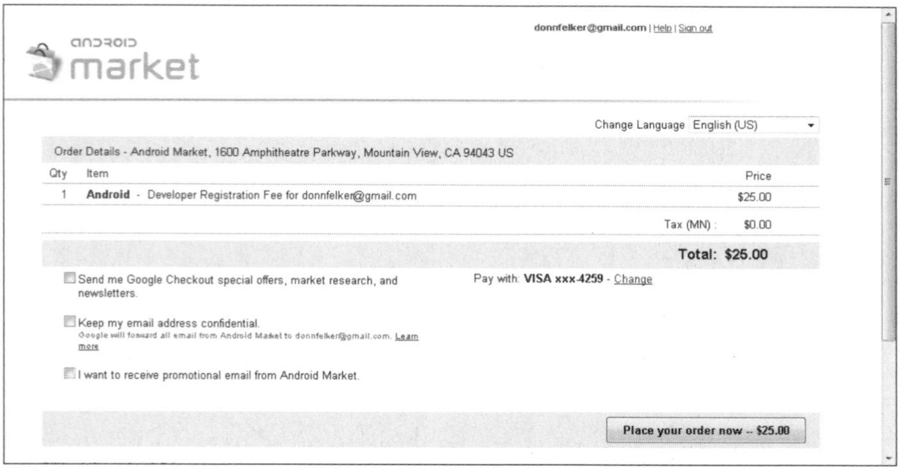

Figure 10-11:
Order
confirmation.

Once complete, you see the "You're Done!" page. (See Figure 10-12.)

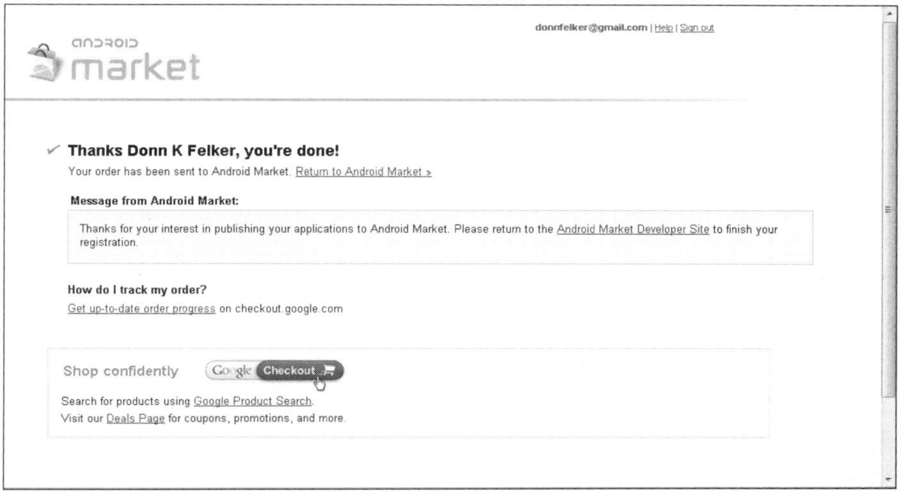

Figure 10-12:
Completing
the
registered
Android
application.

8. **Click the Android Market Developer Site link to arrive at the Android Market Developer Distribution Agreement page, shown in Figure 10-13.**

Figure 10-13:
The terms
agreement.

9. **Click the check box to agree to the terms and click the I Agree, Continue link.**

Once complete, you will now be an Android developer! You are taken to the Android Developer home page. (See Figure 10-14.)

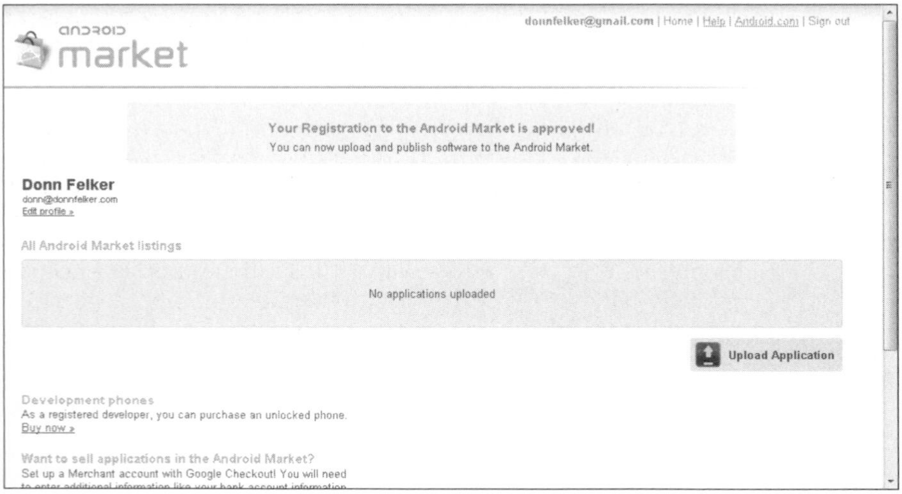

Figure 10-14:
The Android
Developer
home page.

Pricing Your Application

You've got your APK, you're a registered Android developer, and now you're ready to get your app into users' hands, finally. But there is one last really important question that you need to ask yourself — is my app a free app or a paid app?

This decision depends on your application and your needs. If your application is a paid application, you have to determine a price point. Although I cannot decide this for you, I recommend looking at similar applications in the Market to help determine a pricing strategy. Most apps seem to range from $0.99 to $9.99. I rarely see an app over the $10 threshold. Although it may be tempting to price your app high, you're more likely to gain users by choosing the lowest price you can. Keeping your pricing competitive with other similar apps should improve your app's appeal in the market.

The paid-vs.-free argument is an evergreen debate among developers. I've published both paid and free apps, and I have found that either approach can be profitable — you just have to figure out what works best for your application.

If you go with a paid model, that means you start getting money in your pocket within 24 hours of the first sale (barring holidays and weekends) — in that case you'd then receive funds the following business day. However, in my experience, paid applications do not receive many active installs. You are your own marketing team for your app. Unless you've heavily promoted your app, few people will know about your app, and if no one knows about it, then how are they going to know to buy it? Free apps face the same problem, but because users risk so little, your free app will receive a number of active installs anyway.

When a user buys a paid app, he gets a free 15-minute trial period. (It used to be 24 hours.) When a user purchases an app and installs it, Google Checkout authorizes their credit card on file, but the charge remains as an authorization state until 15 minutes after the original purchase time. You can monitor this in your Google Checkout panel. (See the section, "Setting Up a Google Checkout Merchant Account.") During those 15 minutes, the user can use the fully functional application, and if they decide that they do not like the application, they can uninstall it and get a full refund. This is very useful to end users because they do not feel that they are getting penalized a fee for trying your app and not liking it. If they do not uninstall the app and get a refund within 15 minutes, the credit card authorization will turn into a charge and you will receive the funds the following day.

The 15-minute time limit is a blessing to developers because they are more likely to receive funds from purchases from the apps. However, the 15-minute time limit can also be a downfall if your app needs to download additional resources from an online site. A perfect example is a game that only takes up 1 MB of install space but must download game resources (3D objects, sounds, and maps, for example) to your SD Card before you can play. Depending on the user's connection speed), the download could take longer than 15 minutes to complete, and therefore could lock the user into either purchasing the product or opting out and getting a refund before the download completes.

If you choose to go the free route, then users can install the application free of charge. From my experience anywhere from 50 to 80 percent of the users who install your free app will keep the application on the device, while the others uninstall it. The elephant in the room at this point is — How do you make money with free apps?

As the age-old saying goes, nothing in life is free — and the same goes for making money on free apps. When it comes down to it, it's fairly simple — advertising. Mobile advertising agencies can provide you with a third-party library to display ads on your mobile application. The top mobile advertising agencies at this time are Google AdSense, AdMob (which was recently acquired by Google), and Quattro Wireless (recently acquired by Apple). Obtaining a free account from one of these companies is a straightforward process. They offer great SDKs and walk you through how to get ads running on your native Android application. Most of these companies pay on a NET 60-day cycle, so it will be a few months before you receive your first check.

The final method of receiving payment in your applications is through in-app billing. In-app billing is a method in which your application can bill users through the use of the app. The classic example of in-app billing in action is a game that allows users to purchase upgrades (new weapons, magic points, armor, and so on) with real money. The in-app billing component is a large topic and could warrant its own chapter. If in-app billing interests you, please visit the online documentation here: `http://d.android.com/guide/market/billing/index.html`.

Setting Up a Google Checkout Merchant Account

In order to have a paid application on the Android Market, you must set up a Google Checkout Merchant Account. To do this, navigate your browser to `checkout.google.com`, type your Google account username and password, and answer the questions that follow. You provide:

- Personal and business information
- Tax identity information, personal or corporation
- Expected monthly revenue (one billion dollars, right?)

After you have set up a Google Checkout Merchant Account, you can sell your applications.

Getting Screenshots for Your Application

Screenshots are a very important part of the Android Market ecosystem because they allow users to preview your application before installing it. Apps with screenshots have higher install rates than apps without. Allowing users to view a couple running shots of your application can be the determining factor of whether or not a user will install your application. Imagine if you created a game and wanted users to play it. If you spent weeks (or months, for that matter) creating detailed graphics, you'd want the potential users/buyers of the game to see them so that they can see how great your app looks.

In order to grab real-time shots of your application, you need an emulator or physical Android device. To grab the screenshots, perform the following:

1. **Open the emulator and place the widget onto the Home screen.**

2. **In Eclipse, open the DDMS Perspective.**

3. **Choose the emulator in the Devices panel.**

4. **Now click the screenshot button, as shown in Figure 10-15 (the icon is at the top right of the Devices view in Eclipse).**

   Doing this captures a screenshot, as shown in Figure 10-15.

You can make changes on the emulator or device and refresh the screenshot dialog, as shown in Figure 10-15. After this screenshot is taken, you can publish it to the Android Market.

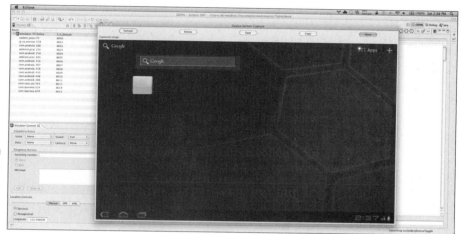

Figure 10-15: A screenshot taken with DDMS.

Uploading Your Application to the Android Marketplace

You've finally reached the apex of the Android application development — the publishing of your application. Don't worry. It's very easy — just follow these steps:

1. **From the Android developer home page (shown in Figure 10-14), click Upload Application.**

   The Upload an Application screen appears.

2. **For the Application .apk file, choose the .apk file that you created earlier in the chapter and then click Upload.**

3. **In the Screenshots section, add two screenshots of your application.**

   To see how to take these shots, see the section, "Getting Screenshots for Your Application," earlier in this chapter. The sizes of these screen shots need to be 320w x 480h, 480w x 800h, 480w x 854h, or 1280w x 800h (please note, Google may add new screen sizes in the future that may not be listed here). Screenshots are not required to publish the app. You should also add a large resolution icon as well. This icon is 512px by 512px. The icon shows up on the online version of the Market: `http://market.android.com`.

4. **Add a promo shot.**

    This promo shot will need to be created in the dimensions of 180w x 120h and should be created in an image-editing program. The promo shot is used for random promotions that Android chooses to showcase when browsing the market. A promo shot is not required to publish the app.

5. **Set the title of your application.**

    I chose Screen Brightness Toggle Widget. This text is indexed for the Android Market search.

6. **Set the description for your application.**

    This is the description that the users see when they inspect your application to determine if they want to install it or not. All of this text is indexed for the Android Market search.

7. **Set the application type.**

    For this app, I set the type to Applications.

8. **Set the Category for the app.**

    I chose Productivity for the Screen Brightness Toggle application because the app is a productivity enhancer.

9. **Select your copy protection.**

    I choose Off — always. When you choose On, the file footprint on the device is usually doubled. If your app is 2 MB in size and you turn on copy protection, your new file footprint when installed on the device will be around 4 MB. Older devices, prior to Android 2.2, could not install applications to the SD Card, therefore internal space was limited; when users ran out of space, they would uninstall the heavyweight apps first to free up the most space. If your app is very heavy weight it's a very good possibility that it will be removed to save space. Keeping the file size small and leaving copy protection set to Off keeps you out of the crosshairs in this issue.

10. **Select the Content rating.**

    This section allows you to specify how you rate your application's content. If you have built an application that is based on, say, adult comedy, you'd want to select High Maturity, whereas your Screen Brightness Toggle, because it's meant for all audiences, should be set to Low Maturity. Think of the content rating for Android apps as the equivalent of movie ratings you see for movies.

11. **Select the list of locations the application should be visible in.**

For example, if your application is an Italian application, deselect All Locations and select Italy as the destination location. This ensures that only devices in the Italy region will see this in the Market. If you leave All Locations enabled, all locations will be able to see your app in the Market.

12. **Fill out the website and e-mail address (and phone, if you'd like).**

    I never provide my phone number because, well, users will call you! Yes, they will call at midnight asking you questions, giving feedback, and so on. I prefer to speak to customers via e-mail as it suits my lifestyle better. If you are writing an app for a different company, yet publishing it under your developer account, you can change the website, e-mail, and phone number so that users do not contact you.

13. **Check that your application meets the Android Content Guidelines and that you complied with applicable laws. Once complete, you have three options:**

    - Publish: Saves and publishes the app to the Market in real time.

    - Save: Saves the changes made, but does not publish the app.

    - Delete: This deletes all the work up until now. Don't do this.

14. **Click Save.**

    This saves your application and returns you to the Android Developer home page, as shown in Figure 10-16.

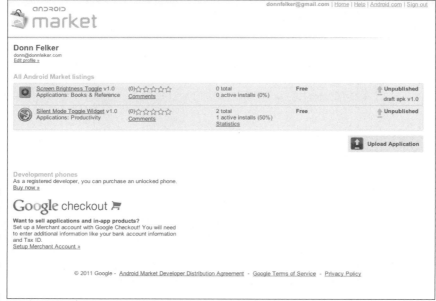

Figure 10-16: The saved app in your Android Developer home page.

On the right side of the screen, notice that the icon states the app is in a saved state. You can use this as a staging area until you're ready to release your app.

15. **When you're ready to release the app, click the title of the app on the Android Developer home page (as shown in Figure 10-16), which opens the Upload an Application screen.**

16. **Scroll to the bottom and click Publish.**

Doing this publishes your application to the Android Market. Figure 10-17 shows the application I just built running in the Android Market on my Motorola XOOM device. I opened the Android Market, navigated to Apps➪Productivity, and clicked the Just In tab, which identifies the apps that have just been released.

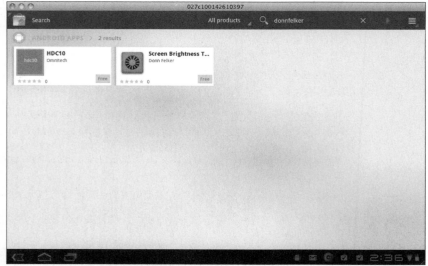

Figure 10-17:
The released application in the Android Market.

You may have noticed one bonus of this process — there is no app approval process as there is with other mobile carriers! You can create an app, right now, publish it, and users will be able to install it right away. Therefore, you can perform a quick release cycle and get new features out the door as quickly as you can get them done, which is very cool.

Clicking Publish and Watching the Installs Soar

You've finally published your first application — now it's time to watch the millions start rolling in, right? Well, kind of. Whether you're an independent developer or a corporate developer, you need to be aware of the end user's experience with your app on various devices. There are various ways to identify how your application is doing:

- ✔ **Current Overall User Rating:** A five-star rating is the best. The higher average rating you have, the better.

- ✔ **User Comments:** Read them! People take the time to write them, so provide them the courtesy of reading them — you'd be surprised the great ideas that people provide to you for free. Most of the time I've found if I implement the most commonly requested feature, users get extremely excited about it and will come back and update their comment with a much more positive boost in rating.

- ✔ **Error Reports:** These reports are submitted by users who were gracious enough to let you know that your app experienced a run-time exception. The error report contains *stack trace* information — that is, a report of the sequence of instructions your app followed that led to the error — which can be very helpful in debugging. (Stack traces are available only for devices running Android 2.2 and later.) Review these reports and fix the errors. Apps that get a lot of force-close errors tend to receive bad reviews — quickly.

- ✔ **Total Installs:** The total number of times your app has been installed.

- ✔ **Active Installs:** The total number of users who currently have your application installed.

  Keep track of total installs versus active installs. Although this isn't the best metric for identifying user satisfaction, it is an unscientific way to determine if users who install your app tend to keep it on their phone. If users are keeping your app, they must like it for one reason or another!

- ✔ **Direct E-mail:** Users can return to the Android Market to find your e-mail address or website address. They will e-mail you to ask you questions about features or to send comments about their user experience. They may also send you ideas about how to improve your app or to ask you to create another app that does something they cannot find on the Market. People love to be a part of the process. Try to reply to users within 24 hours — I aim for four hours — to keep your users happy. Although this is difficult to sustain if your app has a million active users, your customers will appreciate your efforts.

Keeping in touch with your user base is a large task, but by doing so you reap the rewards of dedicated, happy customers who will refer their friends and family to your application.

Part III

Creating a Feature-Rich Application

The 5th Wave   By Rich Tennant

CA CURSOR ASSOCIATES

"Having achieved market dominance in the workstation cursor market, we're excited to introduce our line of home cursors called, MyCursor, available in these exciting colors."

In this part . . .

In Part III, I demonstrate how you can build a feature-rich application. In these chapters I don't trudge through every detail as I do in Part II, but I expand on them. I also discuss a few advanced topics to help you bridge the gap between beginner and advanced Android developer.

In this part, I showcase how and why you would create certain features to enhance users' experiences with your application. At the end of Part III, you will have a fully-functioning advanced application that interacts with a local database and custom preferences.

Chapter 11

Designing the Task Reminder Application

*B*uilding Android applications is fun, but building truly in-depth applications is exciting because you dive into the real guts of the Android
platform. In this chapter, I introduce you to such an application — the Task
Reminder application, which you build from beginning to end over the next
couple of chapters.

In this chapter, you design the app and create the screens and basic processes. Over the course of the next few chapters, you develop these processes to create your app.

This app requires a lot of interaction with the user and with the Android
system, and by developing it, you are introduced to a number of facets of
Android development that can help you in your career. I wish I would have
known some of these things when I started — it would have saved me a lot
of time!

Reviewing the Basic Requirements

The idea behind the Task Reminder application is to remind users through
the use of an alarm system that a task must be performed. It will allow users
to create a list of time-sensitive tasks, and to set a date and time for the
app to remind them of these tasks. To fulfill what is expected of it, the Task
Reminder application has a few basic requirements:

✔ It must be able to accept user input — having a personalized task application that does not allow user input would be silly!

✔ Each task must have a reminder date and time in which the user will be reminded of the task.

✔ The user must be notified of the task when the reminder time has arrived.

✔ Users must be able to delete tasks.

✔ Users must be able to not only add tasks but to edit them.

That's alarming! Scheduling a reminder script

For the Task Reminder application to truly work, you need to implement some sort of reminder-based system. As a developer, the first thing that comes to mind is a scheduled task or `cron` job — methods to handle the scheduled execution of code or scripts in the Windows operating system and the UNIX/Linux operating systems, respectively. Although Android is running the Linux 2.6 kernel, it does not have `cron`. However, it has the `AlarmManager` class, which achieves the same thing. The `AlarmManager` class allows you to specify when your application should start in the future. Alarms can be set as a single-use alarm or as a repeating alarm. You will set the alarm as a single use alarm. This should be perfect for your Task Reminder application.

Storing data

You are exposed to many new features and tools in this application. You may be wondering where to store the activities, the task data, the alarms, and so on. You can choose from the following locations:

✔ **Activities and broadcast receivers:** In one Java package

✔ **Task data:** SQLite database

✔ **Alarm info:** Pulled from the SQLite database and placed in the `AlarmManager` via the intent system

Distracting the user (nicely)

After an alarm fires, you need to notify the user. Although Android provides mechanisms to bring your activity to the foreground when an alarm fires, this is probably not the best approach. Interrupting the user in this way may cause the user to be irritated or confused because an activity started that he did not initiate. However, you can grab the user's attention in other ways, including the following:

- ✔ **Toasts:** A *toast* is a small view that contains a quick message for the user. This message does not persist because it is usually available for only a few seconds at most. A toast never receives focus. I won't use a toast for reminding the user, but instead I use a toast to notify the user when her activity has been saved so that she knows something happened.

- ✔ **Notification Manager:** The `NotificationManager` class is used to notify a user that an event or events have taken place. These events can be placed in the status bar, which is located at the top of the screen. The notification items can contain various views and are identified by icons that you provide. The user can slide the screen down to view the notification.

The best approach, then, to handle the alarms for the Task Reminder application, is the `NotificationManager` class.

Creating the Application's Screens

The first part of the process of developing the Task Reminder application is to create the screens. The Task Reminder app will have two different screens that perform all the basic CRUD (Create, Read, Update, and Delete) functions, as follows:

- ✔ **A task list screen:** A list view that lists all the current tasks in the application by name. This view also allows users to delete a task by long-pressing an item.

- ✔ **An add/edit screen:** A view that allows users to view (Read), add (Create), or edit (Update) a task.

Each screen eventually interacts with a database for changes to be persisted over the long-term use of the application.

Starting the new project

To get started, open Eclipse and create a new Android project with a `Build Target` of Android 3.0 and a `MinSDKVersion` of 11. Provide it with a valid name, package, and activity. The settings I have chosen are shown in Table 11-1. If you prefer, you can open the example Android project for Chapter 11 from this book's companion website. This provides you with a starting point that has the same settings as my project.

Table 11-1	New Project Settings
Property	*Value*
Project Name	Task Reminder
Build Target	Android 3.0 (API Level 11) or greater
Application Name	Task Reminder
Package Name	com.dummies.android. taskreminder
Create Activity	ReminderListActivity
Min SDK Version	11

Note the Create Activity property value — ReminderListActivity. (I chose this name over ListActivity because this is the class we must extend.) Normally I give the first activity in an application the name of MainActivity; however, the first screen that the user will see is a list of current tasks. Therefore, this activity is actually an instance of a ListActivity — hence the name ReminderListActivity.

Creating the task list screen

When working with ListActivity classes (classes responsible for working with lists in Android), I like to have my layout filename contain the word *list,* which makes it easy to find when I open the res/layout directory. I'm going to rename the main.xml file located in the res/layout directory to reminder_list.xml. To rename the file in Eclipse, either right-click the file and choose Refactor⟹Rename or select the file and press Shift+Alt+R.

After you change the filename, you need to update the name of the file in the setContentView() call inside the ReminderListActivity.java file, so your Java code is directed to the right file. Open the file and change the reference to the new filename you chose.

Also, the ReminderListActivity class should be changed so that it inherits from the ListActivity class instead of from the regular superclass activity (in this case ListActivity). Make that change as well.

When you make these changes, your new ReminderListActivity class should look something like Listing 11-1.

Listing 11-1:   The `ReminderListActivity` Class

```
public class ReminderListActivity extends ListActivity {
    /** Called when the activity is first created. */
    @Override
    public void onCreate(Bundle savedInstanceState) {
        super.onCreate(savedInstanceState);
        setContentView(R.layout.reminder_list);
    }
}
```

Now your `ReminderListActivity` references the `reminder_list` layout resource that currently contains the default code generated when you created the project. However, to work with a `ListActivity`, you need to update this layout with new code, as shown in Listing 11-2.

Listing 11-2:   The `reminder_list.xml` Contents

```
<?xml version="1.0" encoding="utf-8"?>
<LinearLayout xmlns:android="http://schemas.android.com/apk/res/android"
    android:layout_width="wrap_content"
    android:layout_height="wrap_content">
    <ListView android:id="@+id/android:list"                          →5
        android:layout_width="fill_parent"
        android:layout_height="fill_parent"/>
        <TextView android:id="@+id/android:empty"                     →8
        android:layout_width="wrap_content"
        android:layout_height="wrap_content"
        android:text="@string/no_reminders"/>                         →11
</LinearLayout>
```

This code is briefly explained as follows:

→5   This line defines a `ListView`, which is an Android view that is used to show a list of vertically scrolling items. The ID of the `ListView` *must* be `@id/android:list` or `@+id/android:list`.

→8   This line defines the empty state of the list. If the list is empty, this view is shown. When this view is present, the `ListView` is hidden because there is no data to display. This view must have an ID of `@id/android:empty` or `@+id/android:empty`.

→11   This line uses a string resource called `no_reminders` to inform the user that no reminders are currently in the system. You need to add a new string resource to the `res/values/strings.xml` file with the name of `no_reminders`. The value I'm choosing is "No Reminders Yet."

Creating the add/edit screen

The Task Reminder application needs one more screen that allows the user to edit a task and its information. This screen will be all-inclusive, meaning that one single activity can allow users to create, read, and update tasks.

In Eclipse, create a new activity that can handle these roles. I'm choosing to call mine `ReminderEditActivity` by right-clicking the package name in the `src` folder and choosing New⇨Class or by pressing Shift+Alt+N and then choosing Class. In the new Java class window, set the superclass to `android.app.Activity` and choose Finish.

A blank activity class now opens. Inside this class, type the following lines that are boldface:

```
public class ReminderEditActivity extends Activity {
    @Override
    protected void onCreate(Bundle savedInstanceState) {
        super.onCreate(savedInstanceState);
        setContentView(R.layout.reminder_edit);                    →5
    }
}
```

In line 5 of the preceding code, I set the layout of the activity to the `reminder_edit` resource, which is defined later in this section. This layout contains the various fields of the task in which the user can edit or create.

You also need to inform the Android platform about the existence of this activity by adding it to the Android Manifest. You can do so by adding it to the `Application` element of the `AndroidManifest.xml` file, as shown here in boldface:

```
<application android:icon="@drawable/icon" android:label="@string/app_name">
    <activity android:name=".ReminderListActivity"
            android:label="@string/app_name">
        <intent-filter>
            <action android:name="android.intent.action.MAIN" />
            <category android:name="android.intent.category.LAUNCHER" />
        </intent-filter>
    </activity>
    <activity android:name=".ReminderEditActivity"
            android:label="@string/app_name" />
</application>
```

If you do not add the activity to the `AndroidManifest.xml` file, you will receive a run-time exception informing you that Android cannot find the class (the activity).

When creating the layout for the adding and editing activity, you need only a few fields, as follows:

- ✓ **Title:** The title of the task as it will appear in the list view.
- ✓ **Body:** The body of the task is where the user types in the details.
- ✓ **Reminder Date:** The date on which the user should be reminded of the task.
- ✓ **Reminder Time:** The time at which the user should be reminded on the reminder date.

When complete and running on a device or emulator, the add/edit screen looks like Figure 11-1.

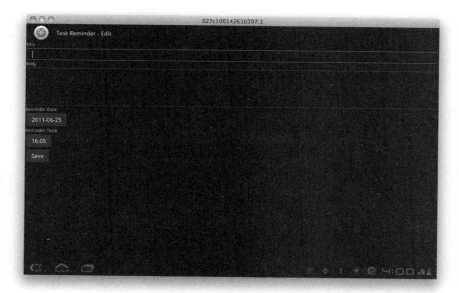

Figure 11-1:
The Add/
Edit Task
Reminder
screen.

To create this layout, create a layout file in the `res/layout` directory with an appropriate name — I'm using `reminder_edit.xml`. To create this file, perform the following steps:

1. **Right-click the** `res/layout` **directory and choose New⇨Android XML File.**

2. **Provide the name in the File field.**

3. **Leave the default type of resource selected — Layout.**

4. **Leave the folder set to** `res/layout`.

5. **Set the root element to** `ScrollView`.

6. **Click Finish.**

You now need to provide all the view definitions to build the screen that you see in Figure 11-1. To do this, type the code shown in Listing 11-3 into your `reminder_edit.xml` file.

Listing 11-3: The `reminder_edit.xml` File

```
<?xml version="1.0" encoding="utf-8"?>
<ScrollView
        xmlns:android="http://schemas.android.com/apk/res/android"
        android:layout_width="fill_parent"
        android:layout_height="fill_parent">                                →5
<LinearLayout                                                             →6
    android:orientation="vertical"                                       →7
    android:layout_width="fill_parent"
    android:layout_height="fill_parent">
    <TextView android:layout_width="wrap_content"
            android:layout_height="wrap_content"
            android:text="@string/title" />                              →12
    <EditText android:id="@+id/title"
            android:layout_width="fill_parent"
              android:layout_height="wrap_content" />                    →15
    <TextView android:layout_width="wrap_content"
            android:layout_height="wrap_content"
            android:text="@string/body" />                               →18
    <EditText android:id="@+id/body"
            android:layout_width="fill_parent"
            android:layout_height="wrap_content"
            android:minLines="5"
            android:scrollbars="vertical"
            android:gravity="top" />                                     →24
    <TextView android:layout_width="wrap_content"
            android:layout_height="wrap_content"
            android:text="@string/date" />                               →27
    <Button
            android:id="@+id/reminder_date"
            android:layout_height="wrap_content"
            android:layout_width="wrap_content"/>                        →31
    <TextView android:layout_width="wrap_content"
            android:layout_height="wrap_content"
            android:text="@string/time" />                               →34
    <Button
```

```
              android:id="@+id/reminder_time"
              android:layout_height="wrap_content"
              android:layout_width="wrap_content" />                    →38
    <Button   android:id="@+id/confirm"
              android:text="@string/confirm"
              android:layout_width="wrap_content"
              android:layout_height="wrap_content" />                   →42
</LinearLayout>
</ScrollView>
```

A brief explanation of the code in Listing 11-3 is as follows:

→**5**   The parent view is a `ScrollView`, which creates a scroll bar and allows the view to be scrolled when the contents of the view are too big for the screen. The screen shown in Figure 11-1 is shown in portrait mode. However, if the device is rotated 90 degrees, the view flips and over half of the view is cut off. The parent `ScrollView` allows the remaining contents of the screen to be scrollable (if there is enough content to fill the screen in a tablet, you'd need well over 15 different tasks listed to initiate this behavior). Therefore, the user can fling his finger upward on the screen to scroll the contents up and see the remainder of the view.

→**6**   A `ScrollView` can only have one child — in this case, it's the main `LinearLayout` that houses the rest of the layout.

→**7**   The orientation of the linear layout is set to vertical to signify that the views inside this layout should be stacked on top of one another.

→**12**   The label for the `Title` field. Please add a string resource with the name `title` and the value of `Title`.

→**15**   The `EditText` that allows the user to provide a title for the task.

→**18**   The label for the `Body` field. Please add a string resource with the name `body` and the value of `Body`.

→**24**   The `EditText` that defines the `Body` field. The `EditText` view has set the `minLines` property to 5 and the `gravity` property to `top`. This informs the Android platform that the `EditText` is at least five lines tall, and when the user starts typing, the text should be bound to the top of the view (the gravity).

→**27**   The reminder date label. This label also uses a string resource. You need to add a string resource with the name of `"date"` and a value of `"Reminder Date"`.

→**31**   The reminder date button. When this button is clicked, a `DatePickerDialog` is launched — this allows the user to choose a date with a built-in Android date picker. When the date is set via the `DatePicker`, the value of the date is set as the button text.

→**34**   The reminder time label. This label uses a string resource. You will need to add a string resource with the name of `"time"` and a value of `"Time"`.

→**38**   The time reminder button. When this button is clicked, a `TimePicker` is launched — this allows the user to choose a time with a built-in Android time picker. When the time is set via the `TimePickerDialog`, the value of the time is set as the button text.

→**42**   The confirmation button, which, when clicked, will save the values of the form. For the text value, please add a string resource with the name `confirm` and the value of `Save`.

Creating Your First List Activity

In order for a user of your application to view his tasks, you must create a list of these tasks in Android. Lists in Android are handled by the `ListView` and `ListActivity`. In the section below, you create a list that houses the list of tasks the user has created by using the `ListView` and `ListActivity`. You also create a custom row that each one of the task items will use in the list to display itself.

The `ListActivity` class displays a list of items by binding to a data source, such as an array or cursor, and exposes callback methods that get executed when the user selects an item. However, to build a list of items to display in a list, you need to add a layout that defines what each row will look like.

A cursor provides random read and write access to the result set that is returned by a database query.

Add a new layout to the `res/layout` directory with a root element of `TextView` and give it a proper name for a row type of item — something like `reminder_row.xml`. Inside this view, type the code shown in Listing 11-4.

Listing 11-4:   The `reminder_row.xml` File

```
<?xml version="1.0" encoding="utf-8"?>
<TextView
    xmlns:android="http://schemas.android.com/apk/res/android"
    android:id="@+id/text1"
    android:layout_width="fill_parent"
    android:layout_height="fill_parent"
    android:padding="10dip"/>
```
→**4**

This code simply defines a row in which text values can be placed with a padding of ten density-independent pixels. Line 4 defines the ID of the view that you will reference when loading the list with data.

The view you just added is actually provided out of the box in the Android system. If you look at the Android documentation for `simple_list_item_1` under `Android.R.layout` and inspect it via the Android source control repository, you will see virtually the same XML definition.

The `ListActivity` requires that an adapter fill the contents of the list view. Various adapters are available, but because you have not built a data store yet (you build it with an SQLite database in Chapter 14), you should create fake data so that you can see the list in action. When you have the fake data, you should set the `ListActivity`'s adapter with a call to `setList-Adapter()`. You do this next.

Getting stubby with fake data

Inside the `onCreate()` method of the `ReminderListActivity.java` file, after the call to `setContentView()`, add the following code:

```
String[] items = new String[] { "Say Hi to Gina", "Go Crazy", "Make a Bazillion
            Dollars", "Plan Vacation" };                                →1

ArrayAdapter<String> adapter =
   new ArrayAdapter<String>(this, R.layout.reminder_row, R.id.text1,
                        items);                                         →4
setListAdapter(adapter);                                               →5
```

A brief explanation of the code is as follows:

→1   This line creates an array of string items. These items eventually will be displayed in the list.

→4   This line creates a new `ArrayAdapter` of string types. An `ArrayAdapter` manages a `ListView` backed by an arbitrary number of arbitrary objects — in this case, a simple string array. This code uses Java generics, which allow the developer to specify the type of object that the `ArrayAdapter` will be working with. The constructor of the `ArrayAdapter` contains the following:

• `this`: The current context (Because the activity is an implementation of the `Context` class, I can use the current instance as the context.)

- R.layout.reminder_row: The row layout that should be used for each row in the ListView

- R.id.text1: The ID of the TextView inside R.layout. reminder_row in which to place the values from the array

- items: The array of strings to load into the ListView

→5 The call to setListAdapter() that informs the ListActivity how to fill the ListView. In this case, I am using the ArrayAdapter created on line 4 to load the ListView.

Start the Android application by choosing Run⇨Run or by pressing Ctrl+F11, and you should see a screen similar to the one in Figure 11-2.

Figure 11-2:
The Task
Reminder
running
with fake/
stubbed
data.

Handling user click events

The items in the list expose click events that allow the user to interact with each item. Android View objects have two main types of click events:

- ✔ **Click:** The user taps a view such as a button.
- ✔ **Long click:** The user taps and holds his finger on a button for a few moments.

Each view or activity can intercept these events through various methods. In the following section, I show you how to respond to each type of event in a `ListActivity`. In Chapter 13, I demonstrate responding to `Button` click events.

Dealing with (regular) clicks

The `ListActivity` in Android does a lot of the event-handling heavy lifting for you — which is good because programming shouldn't be a physical exercise!

To handle a click in a `ListActivity`, then, just add the following method after the `onCreate()` method in the `ReminderListActivity.java` file:

```
@Override
protected void onListItemClick(ListView l, View v, int position, long id) {
    super.onListItemClick(l, v, position, id);
}
```

This code overrides the default implementation of `onListItemClick()` that is provided by the `ListActivity`. Now when a list item is clicked, this method is called and the following parameters are passed into the call:

- ✔ `l`: The `ListView` where the click happened
- ✔ `v`: The item that was clicked with the `ListView`
- ✔ `position`: The position of the clicked item in the list
- ✔ `id`: The row ID of the item that was clicked

Using these variables, you can determine which item was clicked and then perform an action based on that information. The plan is that when an item is clicked, an intent that opens the `ReminderEditActivity` gets fired, which allows users to edit the item, as shown in the section "Starting new activities with intents," later in this chapter.

Dealing with long clicks

Long clicks, also known as *long presses,* occur when a user presses a view for an extended period of time. To handle the list item's long-click event in a `ListActivity`, add the following line of code at the end of the `onCreate()` method:

```
registerForContextMenu(getListView());
```

The outer method, `registerForContextMenu()`, is responsible for registering a context menu to be shown for a given view. (However, multiple views can show a context menu; it's not just limited to a single view.) Therefore, each list item is eligible to create a context menu. The `registerForContextMenu()` accepts a `View` object as a parameter that the `ListActivity` should register as eligible for the context menu creation. The inner method, `getListView()`, returns a `ListView` object that is used for the registration. The call, `getListView()`, is a member of the `ListActivity` class.

Now that you've registered the `ListView` to be eligible to create a context menu, you need to respond to the long-click event on any given item. When an item is long-clicked in the `ListView`, the `registerForContextMenu()` recognizes this and calls the `onCreateContextMenu()` method when the context menu is ready to be created. In this method, you set up your context menu.

At the end of the `ReminderListActivity.java` class file, type the following method:

```
@Override
public void onCreateContextMenu(ContextMenu menu, View v, ContextMenuInfo
              menuInfo) {
    super.onCreateContextMenu(menu, v, menuInfo);
}
```

This method is called with the following parameters:

- ✔ `menu`: The context menu that is being built.
- ✔ `v`: The view for which the context is being built (the view you long-clicked on).
- ✔ `menuInfo`: Extra information about the item for which the context menu should be shown. This can vary depending on the type of view in the `v` parameter.

Inside this method, you can modify the menu that will be presented to the user. For example, when a user long-presses an item in the task list, you want to allow her to delete it. Therefore, you need to present her with a Delete context menu option. You add the Delete item to the context menu in Chapter 13.

Identifying Your Intent

Though some applications require only a couple of screens (such as the Task Reminder application which requires a way to view the task and edit a task), a lot can be happening behind the scenes. One such notable interaction is the introduction of new screens as the user uses various features of the application. As with any application with a rich feature set, the user can interact with each screen independently. Therefore the big question arises: "How do I open another screen?"

Screen interaction is handled through Android's intent system. I have covered the intent system in detail in Chapter 8, but I have not covered an example of how to navigate from one screen to the next using an intent. Thankfully, it's a simple process — and I bet you're happy about that!

Starting new activities with intents

Activities are initiated through the Android intent framework. An `Intent` is a class that represents a message that is placed on the Android intent system (similar to a message-bus type of architecture), and whoever can respond to the intent lets the Android platform know, resulting in either an activity starting or a list appearing of applications to choose from (this is known as a chooser, explained shortly). One of the best ways to think of an intent is to think of it as an abstract description of an operation. (For more on intents, refer to Chapter 8.)

Starting a particular activity is easy. In your `ReminderListActivity`, type the following code into the `onListItemClick()` method:

```
@Override
protected void onListItemClick(ListView l, View v, int position, long id) {
    super.onListItemClick(l, v, position, id);
    Intent i = new Intent(this, ReminderEditActivity.class);        →4
    i.putExtra("RowId", id);                                        →5
    startActivity(i);                                               →6
}
```

A brief explanation of each line is as follows:

→4   This line creates a new intent by using the `Intent` constructor that accepts the current context, which is `this` (the current running activity), as well as a class that the Intent system should attempt to start — the Reminder Edit activity.

→5   This line places some extra data into the `Intent` object. In this instance, I'm placing a key/value pair into the intent. The key is `RowId`, and the value is the ID of the view that was clicked. This value is placed into the intent so that the receiving activity (the `ReminderEditActivity`) can pull this data from the `Intent` object and use it to load the information about the intent. Right now, I'm providing fake/stub data; therefore, nothing displays. However, after Chapter 14, you see data flowing into the `ReminderEditActivity`.

→6   This line starts the activity from within the current activity. This call places the intent message onto the Android intent system and allows Android to decide how to open that screen for the user.

Retrieving values from previous activities

Sometimes, depending on your activity's needs, no extra data is passed when an activity starts. The activity starts, runs, and that's the end of it. However, in some instances an activity needs to be able to pull data out of the incoming intent to figure out what to do. For your Reminder Edit activity, you provided some extra data with the initiating intent. This is the `RowId`. (See the section "Starting new activities with intents," earlier in this chapter.) In Chapter 14, you use this `RowId` on the `ReminderEditActivity` to pull the data from the SQLite database and display it to the user.

To pull the data out of an incoming intent, type the following at the end of the destination activity's — in this case, the `ReminderEditActivity`'s — `onCreate()` method:

```
if(getIntent() != null) {                                          →1
    Bundle extras = getIntent().getExtras();                       →2
    int rowId = extras != null ? extras.getInt("RowId") : -1;      →3
    // Do stuff with the row id here
}
```

A brief explanation of each line of code is as follows:

→1   The `getIntent()` method is provided by the `Activity` base class. This method retrieves any incoming intent to the activity. On this line, I am making sure that it is not null so that I know it's safe to work with.

→2   The bundle is retrieved from the intent through the `getExtras()` call. A *bundle* is a simple key/value pair data structure.

→3   On this line, I am using the ternary operator to identify whether the bundle is null. If the bundle is not null, I retrieve the `RowId` that is contained in the intent that was sent from the previous activity through the `getInt()` method. Although I am not doing anything with it in this instance, in Chapter 12, I use this row ID to query the SQLite database to retrieve the `Task` record to edit.

When the SQLite database is in place (which is done in Chapter 14), the record will be retrieved from the database and the various values of the task will be presented to the user onscreen through an editable form so that the user can edit the task.

Creating a chooser

At some point in your Android development career, you will run into a particular instance where you need to provide the user with a list of applications that can handle a particular intent. A common example of this is when you want the user to be able to share some data with a friend via a common networking tool, such as e-mail, SMS, Twitter, Facebook, Google Latitude, and so on.

The Android Intent system was built to handle these types of situations through the use of a *chooser*. Although a chooser isn't necessary for your Task Reminder application, it is often very handy — which is why I include it here. The code to display various available options to the user is shown in Listing 11-5.

Listing 11-5:   Creating an Intent Chooser

```
Intent i = new Intent(Intent.ACTION_SEND);                          →1
i.setType("text/plain");                                            →2
i.putExtra(Intent.EXTRA_TEXT, "Hey Everybody!");                    →3
i.putExtra(Intent.EXTRA_SUBJECT, "My Subject");                     →4
Intent chooser = Intent.createChooser(i, "Who Should Handle this?");→5
startActivity(chooser);                                             →6
```

A brief explanation of each line in Listing 11-5 is as follows:

→1   Creates a new intent that informs the Intent system that you would like to send something — think of this as something you want to mail to another person. You are *intending* to send something to someone else.

→2   Defines the content type of the message — this can be set to any explicit MIME type. MIME types are case-sensitive, unlike RFC MIME types, and should always be typed in lowercase letters. The mime type specifies the type of the intent; therefore, only applications that can respond to this type of intent will show up in the chooser.

→3   Places extra data into the intent — in this case, the body of the message that the application will use. If, say, an e-mail client is chosen, this data will end up as the e-mail body. If Twitter is chosen, it will be the message of the tweet. Each application that responds to the intent can handle this extra data in its own special manner. The data in the destination application may not be handled as you expect. The developer of such an application determines how the application should handle the extra data.

→4   Similar to line 3, but this time the "extra subject" is provided. If an e-mail client responds, this data normally ends up as the subject of the e-mail.

→5   Creates the chooser. The `Intent` object has a static helper method that helps you create a chooser. The chooser is an intent itself. You simply provide the target intent (that is, what you want to happen) as well as a title for the pop-up chooser that is shown.

→6   Starts the intent. This creates the chooser from which the user can choose an application.

The chooser that is created from Listing 11-5 is shown in Figure 11-3.

If the Intent system cannot find any valid applications to handle the intent, the chooser is created with a message informing the user that no applications could perform the action, as shown in Figure 11-4.

Choosers are a great way to increase the interoperability of your application. However, if you simply called `startActivity()` without creating a chooser — such as using `startActivity(i)` instead of `startActivity(chooser)` in Listing 11-5 — your application might crash. The application would crash because Android assumes you know what you're doing. By not including a chooser, then, you're telling Android to assume that the destination device has at least one application to handle the intent. If this is not the case, Android will throw an exception (visible through DDMS) informing you that no class can handle the intent. To the end user, this means your app has crashed.

Figure 11-3:
The new chooser that was created.

To provide a great user experience, always provide an intent chooser when firing off intents meant for interoperability with other applications. It provides the sort of smooth and consistent usability model that Android users expect.

Figure 11-4:
A chooser
informing
the user
that Android
could
not find a
matching
application
to handle
the Intent.

Chapter 12

Going à la Carte with Your Menu

*O*kay, I don't mean that kind of menu. Sure, I wish I were down at my favorite Mexican restaurant, ordering some excellent chips and salsa, but I'm not. Instead, I'm talking about menus inside an Android application!

Android tablets provide a simple mechanism for you to add menus to your applications. You find the following types of menus:

✔ **Action bar:** The action bar is a new addition to Android 3.0. The action bar contains the most commonly used menu options you would encounter in a tablet application. These menu options, located in the upper-right side of the application screen, appear as icons with text or simply as text references to the action that they perform.

✔ **Options menu:** The options menu is the most common type of menu that you will work with because it is the primary menu for an activity. This menu is presented when a user presses the Menu key on the device. Within the options menu are two groups:

   • *Icon:* These menu options are available at the bottom of the screen. The device supports up to six menu items — the only menu items that support the use of icons. They do not support check boxes or radio buttons, however.

   • *Expanded:* This group is a list of menu items beyond the original six menu items present in the Icon menu. When the user places more than six items on the Icon menu, the More menu icon appears onscreen. When you click this icon, the additional menu items appear.

✔ **Context menu:** This menu is a floating list of menu items presented when a user long-presses a view.

✔ **Submenu:** A submenu is a floating list of menu items the user can open by clicking a menu item on the options menu or on a context menu. A submenu item cannot support nested submenus.

In this chapter, you create an options menu as well as a context menu for your Task Reminder app. Feel free to grab the full application source code from this book's companion website if you happen to get lost.

Seeing What Makes a Menu Great

If you have an Android device and you've downloaded a few applications from the Android Market, I'm sure that you've encountered a few bad menu implementations. What does a bad menu implementation look like?

A bad menu is a menu that provides very little (if any) helpful text in the menu description and provides no icon. Common menu faux pas include

- ✔ A menu without an icon
- ✔ A menu item that should be in the action bar yet is not
- ✔ A poor menu title
- ✔ No menu
- ✔ A menu that does not do what it states it will

Although all these issues indicate a bad menu, the two biggest are the first two. This may sound a bit odd, but think about it for a second. If a menu does not have an icon, that means the developer has not taken the time to provide a good user interface — and therefore a good user experience — for the user. A good menu should have a visual as well as a textual appeal. The appearance of a menu icon shows that the developer actually thought through the menu-creation process and took the time to decide which icon best suited the application. Also, failing to support the menu in the action bar reflects a poor design, especially if the menu has items accessed often in the application. If a developer does not place an actionable item into the action bar, it's clear that he has not properly thought through the use of the application.

It's not enough just to avoid these errors, though. Just because an application has menu icons does not mean that the menu is great. Along the same lines, even if an application has action bar items, it does not mean that the items are the correct ones or are the best ones for that app.

Creating Your First Menu

You can create a menu through code or you can create it through an XML file that is provided in the `res/menu` directory. The preferred method of creating menus is to define menus through XML and then *inflate* them into a programmable object with which you can interact. (*Inflate* here refers to the

Android process of turning an XML file into a Java object.) This helps separate the menu definition from the actual application code.

Defining the XML file

To define an XML menu, follow these steps:

1. **Create a** menu **folder in the** res **directory.**

2. **Add a file by the name of** list_menu.xml **to the menu directory.**

3. **Type the code from Listing 12-1 into the** list_menu.xml **file.**

Listing 12-1: Menu for the `ReminderListActivity`

```
<?xml version="1.0" encoding="utf-8"?>
<menu xmlns:android="http://schemas.android.com/apk/res/android">
  <item android:id="@+id/menu_insert"
        android:icon="@android:drawable/ic_menu_add"          →4
        android:title="@string/menu_insert" />                →5
</menu>
```

Notice that a new string resource is included in line 5. You need to create that (which you do in Step 4). The android:icon value in line 4 is a built-in Android icon. You do not have to provide this bitmap in your drawable resources. The ldpi, mdpi, and hdpi versions of this icon are all built into the Android platform. To view other resources available to you, view the android.R.drawable documentation here: http://developer. android.com/reference/android/R.drawable.html.

All resources in the android.R class are available for you to use in your application and are recommended because they give your application a common and consistent user interface and user experience with the Android platform.

4. **Create a new string resource with the name** menu_insert **with the value of "Add Reminder" in the** strings.xml **resource file.**

5. **Open the** ReminderListActivity **class and type the following code into the file:**

```
@Override
public boolean onCreateOptionsMenu(Menu menu) {
    super.onCreateOptionsMenu(menu);
    MenuInflater mi = getMenuInflater();                      →4
    mi.inflate(R.menu.list_menu, menu);
    return true;
}
```

On line 4, I obtain a `MenuInflater` capable of inflating menus from XML resources. After the inflater is obtained, the menu is inflated into an actual menu object on line 5. The existing menu is the menu object that is passed into the `onCreateOptionsMenu()` method.

 6. **Install the application in the emulator and click the Menu button.**

You should see what's shown in Figure 12-1. Notice how there is no menu item? The next step shows you how to get the icon to appear in the action bar.

Figure 12-1:
The Add
Reminder
menu.

 7. **Open the `list_menu.xml` file and type the following code into the file on line 6.**

```xml
<?xml version="1.0" encoding="utf-8"?>
<menu xmlns:android="http://schemas.android.com/apk/res/android">
  <item android:id="@+id/menu_insert"
        android:icon="@android:drawable/ic_menu_add"
        android:title="@string/menu_insert"
        android:showAsAction="ifRoom|withText" />        →6
    </menu>
```

Because adding a new reminder is a common action within this application, you want this menu to be placed in the action bar. Adding the code in this step does this.

 8. **Install the application in the emulator, and the menu item appears in the action bar, as shown in Figure 12-2.**

The Add Reminder menu icon

Figure 12-2:
The Add
Reminder
menu icon
in the
action bar.

Handling user actions

The menu has been created, and now you'd like to perform some type of
action when it is clicked. To do this, type the following code at the end of the
`ReminderListActivity` class file:

```
@Override
public boolean onMenuItemSelected(int featureId, MenuItem item) {          →2
    switch(item.getItemId()) {                                             →3
        case R.id.menu_insert:                                             →4
            createReminder();                                              →5
            return true;                                                   →6
    }

    return super.onMenuItemSelected(featureId, item);
}
```

The lines of code are explained in detail here:

→2   This method is called when a menu item is selected. The `featu-`
      `reId` parameter identifies the panel on which the menu is located.
      The `item` parameter identifies which menu item was clicked.

→3   To determine which item you're working with, compare the ID of the menu items with the known menu items you have. Therefore, a `switch` statement is used to check each possible valid case. You obtain the menu's ID through the `MenuItem` method `getItemId()`.

→4   In this line you use the ID of the menu item defined in Listing 12-1 to see whether that menu item was clicked.

→5   If the Add Reminder menu item was clicked, the application is instructed to create a reminder through the `createReminder()` method (defined in the next section).

→6   This line returns true to inform the `onMenuItemSelected()` method that a menu selection was handled.

You may receive compilation errors at this time, but don't worry! You resolve those errors in the next section.

Creating a reminder task

The `createReminder()` method is used to allow the user to navigate to the `ReminderEditActivity` to create a new task with a reminder. Type the following method at the bottom of your `ReminderListActivity` class file:

```
private static final int ACTIVITY_CREATE=0;
private void createReminder() {
    Intent i = new Intent(this, ReminderEditActivity.class);
    startActivityForResult(i, ACTIVITY_CREATE);                      →4
}
```

This code creates a new intent that starts the `ReminderEditActivity`. The `startActivityForResult()` call on line 4 is used when you would like to know when the called activity is completed. You may want to know when an activity has been returned so that you can perform some type of action. In the case of the Task Reminder application, you would want to know when the `ReminderEditActivity` has returned to repopulate the task list with the newly added reminder. This call contains the following two parameters:

✔ `Intent i`: This intent starts the `ReminderEditActivity`.

✔ `ACTIVITY_CREATE`: This request code is returned to your activity through a call to `onActivityResult()`. The request code in this is a classwide constant, defined at the top of the `ReminderListActivity` as such:

```
private static final int ACTIVITY_CREATE=0;
```

Completing the activity

The final call is the `onActivityResult()` call. When the `ReminderEditActivity` completes, the `onActivityResult()` method is called with a request code, a result code, and an intent that can contain data back to the original calling activity. Type the following code into the bottom of the `ReminderListActivity` class file:

```
@Override
protected void onActivityResult(int requestCode, int resultCode, Intent intent)
{
    super.onActivityResult(requestCode, resultCode, intent);
    // Reload the list here
}
```

At this point, this call does nothing, but it will be used in Chapter 14 when you need to reload the tasks from the SQLite database. These parameters are as follows:

- ✔ **requestCode:** The integer request code that was provided in the original `startActivityForResult()` call. If your activity starts various other child activities with various request codes, this allows you to differentiate each returning call through a switch statement — very similar to the `onMenuItemSelected()` item `switch` statement mechanism.

- ✔ **resultCode:** The integer result code returned by the child activity through its `setResult()` call. The result code allows you to determine whether your requested action was completed, canceled, or terminated for any other reason. These codes are provided by you to determine what happened between activity calls.

- ✔ **intent:** An intent that the child activity can create to return result data to the caller (various data can be attached to intent "extras"). In the example shown, this intent instance is the same one that is passed into the `onActivityResult()` method.

The superclass is called to take care of any extra processing that may need to take place by the Android system as it builds the menu.

Creating a Context Menu

A context menu is created when a user long-presses a view. The context menu is a floating menu that hovers above the current activity and allows users to choose from various options.

The Task Reminder application needs a mechanism by which to delete a task when it is no longer needed in the app. In this section, you implement this feature as a context menu. Users long-press the task in the list, and they receive a context menu that allows them to delete the task when the context menu is clicked.

Thankfully, creating a context menu is quite similar to creating an option menu. The menu can be defined in XML and can be inflated using the same mechanism that is used in the creation of an options menu. To create a context menu, then, you need to call `registerForContextMenu()` with a view as the target. You created one of these in Chapter 11: After it is created, you need to override the `onCreateContextMenu()` call — also demonstrated in Chapter 11.

Creating the menu XML file

To create this menu, create a new XML file in the `res/menu` directory. I'm going to name mine `list_menu_item_longpress.xml`. Type the following into the XML file:

```xml
<?xml version="1.0" encoding="utf-8"?>
<menu
    xmlns:android="http://schemas.android.com/apk/res/android">
    <item android:id="@+id/menu_delete"
            android:title="@string/menu_delete" />
</menu>
```

Notice that the `title` property uses a new string resource `menu_delete`. You need to create a new string resource with the name of `menu_delete` and the value of "Delete Reminder." Also note that I do not have an icon associated with this menu. This is because a context menu does not support icons because context menu items are simply a list of menu options that floats above the current activity.

Loading the menu

To load the menu, type the following code into the `onCreateContext-Menu()` method:

```java
@Override
public void onCreateContextMenu(ContextMenu menu, View v,
            ContextMenuInfo menuInfo) {
    super.onCreateContextMenu(menu, v, menuInfo);
    MenuInflater mi = getMenuInflater();
    mi.inflate(R.menu.list_menu_item_longpress, menu);
}
```

This code performs the same function as the `onCreateOptionsMenu()` call, but this time you are inflating the menu for the context menu and you are loading the context menu. Now, when you long-press a list item in the list view, you receive a context menu, as shown in Figure 12-3.

Figure 12-3:
The context menu in the Task Reminder application.

Handling user selections

Handling the selection of these menu items is very similar to the way you handled it in the option menu. Just type the following code into the bottom of your `ReminderListActivity` file:

```
@Override
public boolean onContextItemSelected(MenuItem item) {          →2
    switch(item.getItemId()) {                                 →3
        case R.id.menu_delete:                                 →4
            // Delete the task
            return true;
    }
    return super.onContextItemSelected(item);
}
```

The code lines are explained here:

→**2** This method is called when a context menu item is selected. The item parameter is the item that was selected in the context menu.

→**3** A switch statement is used to determine which item was clicked based upon the ID as defined in the list_menu_item_long press.xml file.

→**4** This line is the ID for the menu_delete button in the list_ menu_item_longpress.xml file. If this menu option is selected, the following code would perform some action based on that determination. Nothing is happening in this code block in this chapter, but in Chapter 14, I delete the task from the SQLite database.

You can add many different context menu items to the list_menu_item_ longpress.xml file and switch between them in the onContextMenuItem Selected() method call — each performing a different action.

Chapter 13

Handling User Input

*I*t's rare to find an application that does not allow user input. Users almost always need to interact with an application in one way or another — through the use of text, date pickers, time pickers, or any other input mechanisms, such as radio buttons or check boxes,. Although the input mechanism may provide a way for users to interact with your application, they won't be chit-chatting and spurring up small talk. User input, in this sense, refers to buttons, screen-dragging, menus, long-pressing, and various other options. In this chapter, I focus solely on user input in the form of free-form text, date and times, and alerts — the sort of input you're most likely to encounter in your Task Reminder application.

Creating the User Input Interface

The most common input type is free-form text — known as an `EditText` widget. In other programming platforms, this is known as a text box. With an `EditText` widget, you can provide an onscreen keyboard or the user can elect to use the physical keyboard (if the device provides one) to enter input.

Creating an EditText widget

In Chapter 11, you create a view layout XML file with the name of `reminder_edit.xml` that contains the following code:

```
<EditText android:id="@+id/title"
    android:layout_width="fill_parent"
    android:layout_height="wrap_content" />
```

This snippet of code defines the text input for the title of the task. The snippet creates an input on the screen so that the user can type into it. This `EditText` widget spans the entire width of the screen and takes up only as much room as it needs in regard to height. When selected, Android automatically opens the onscreen keyboard to allow the user to enter some input on the screen. This example is minimalistic compared to the following `EditText` example, which is also created in the `reminder_edit.xml` layout file:

```
<EditText android:id="@+id/body" android:layout_width="fill_parent"
    android:layout_height="wrap_content"
    android:minLines="5"
    android:scrollbars="vertical"
    android:gravity="top" />
```

Here, you create the body description text for the task. The layout width and height are the same as in the previous `EditText` widget, but the differences between these two widgets are outlined in the following three properties:

- ✔ `minLines`: This property specifies how tall the `EditText` view should be. The `EditText` view is a subclass of the `TextView` object; therefore, they share the same property. Here you specify that the `EditText` object on the screen be at least five lines tall. This is so that the view resembles a text input that is for long messages. This is much like the body portion of any e-mail client: the height of the body is much larger than the height of the subject. In this case, though, the body is much larger than the title.

- ✔ `scrollbars`: This property defines which scroll bars should be present when the text overflows the available input real estate. In this instance, you specify vertical so that scroll bars appear on the side of the `EditText` view.

- ✔ `gravity`: By default, when the user places focus into an `EditText` field, the text aligns to the middle of the view, as shown in Figure 13-1. However, this is not what users would expect when they work with a multiline input mechanism. The user normally expects the input to have the cursor placed at the top of the `EditText` view. To do this, you must set the gravity of the `EditText` view to "top." This forces the text to gravitate to the top of the `EditText` input.

Figure 13-1:
An EditText
view
without the
gravity set.

Displaying an onscreen keyboard

Because some devices do not have a physical keyboard, an onscreen keyboard must be present to interact with the input mechanisms. The EditText view is very versatile and its properties can be configured many ways. One of these properties is responsible for the display of the onscreen keyboard.

Why would you need to adjust the onscreen keyboard? It's simple: Different EditText input types sometimes need different keys. For example, if the EditText is a phone number, the onscreen keyboard should display numbers only. If the EditText value is an e-mail address, the onscreen keyboard should display common e-mail style attributes — such as the at symbol (@). The way you configure the onscreen keyboard can increase the usability of your application.

You configure the onscreen keyboard through the inputType property on the EditText view. This configuration offers far too many options for me to cover in this book, but if you like, you can review all of them at http://developer.android.com/reference/android/widget/TextView.html#attr_android:inputType.

Getting Choosy with Dates and Times

What would your Task Reminder application be without a way to set the date and time for the reminder? Well, it wouldn't be a Task Reminder application at all! It would simply be a Task List application — and that's kind of boring if you ask me.

If you've done any programming with dates and times in other programming languages, you know that building a mechanism for a user to enter a date and a time can be a painstaking process all in itself. However, I'm happy to report that the Android platform has relieved Android programmers of this difficulty. The Android platform provides two classes that assist you in this process — the DatePicker and TimePicker — and these pickers also provide built-in classes that allow you to pop up a dialog box so that the user can select a date and a time. This means you have the option to either embed the DatePicker or TimePicker into your application's views or to use the Dialog classes, which saves you the process of creating a view in which to contain the DatePicker and TimePicker views.

Enough jibber-jabber about what the picker widgets can do. I'm sure you're ready to start using them!

Creating picker buttons

You have not added the DatePicker or TimePicker to the Task Reminder application yet, but you do so in this section. Part of the reminder_edit. xml file contains mechanisms to help display the DatePicker and TimePicker. These mechanisms are located after the EditText definitions discussed earlier in this chapter — two buttons with two labels above them, as shown in Listing 13-1. Please add this code to your reminder_edit.xml file now.

Listing 13-1:   The Date and Time Buttons with Their Corresponding TextView Labels

```
<TextView android:layout_width="wrap_content"                        →1
    android:layout_height="wrap_content"
    android:text="@string/date" />
<Button                                                              →4
    android:id="@+id/reminder_date"
    android:layout_height="wrap_content"
    android:layout_width="wrap_content"
    />
<TextView android:layout_width="wrap_content"                        →9
    android:layout_height="wrap_content"
    android:text="@string/time" />
```

```
<Button                                                    →12
    android:id="@+id/reminder_time"
    android:layout_height="wrap_content"
    android:layout_width="wrap_content"
    />
```

The code lines are explained here:

→**1**   This line is the `TextView` label for the Date button. This displays the value of `ReminderDate` according to the string resource.

→**4**   This line defines a button that the user clicks to open the `DatePickerDialog`, as explained in the next section, "Wiring up the date picker."

→**9**   This line is the `TextView` label for the Time button. This displays the value of `ReminderTime` according to the string resource.

→**12** This line defines a button that the user clicks to open the `TimePickerDialog`, as explained in the next section, "Wiring up the date picker."

Wiring up the date picker

When the user clicks the Date button, he should be able to edit the date. To do this, you need to declare the `onClickListener()` for the Date button, and then, when a click is detected, open a dialog box that allows the user to make the change. I show you how to do this in the following sections.

Setting up the Date button click listener

To implement this functionality, open the activity where your code is to be placed — for the Task Reminder application, open the `ReminderEditActivity.java` file.

In the `onCreate()` method, type the following code:

```
registerButtonListenersAndSetDefaultText();
```

Eclipse informs you that you need to create the method, so do that now. The easiest way to do this is by hovering your cursor over the method call's squiggly and choosing from the resulting context menu the "Create method 'registerButtonListenersAndSetDefaultText()'" option. In the `registerButtonListenersAndSetDefaultText()` method, type the code shown in Listing 13-2.

Listing 13-2: Implementing the Date Button Click Listener

```
mDateButton.setOnClickListener(new View.OnClickListener() {          →1

    @Override
    public void onClick(View v) {                                    →4
        showDialog(DATE_PICKER_DIALOG);                              →5
    }
});
updateDateButtonText();                                              →8
updateTimeButtonText();                                              →9
```

This code is explained as follows:

→1   This line uses the `mDateButton` variable. As you have probably noticed, you have not defined this variable anywhere. You need to define this variable at the top of the class file. After this variable is defined, you can set the `onClickListener()` for the button. The `onClickListener()` is what executes when the button is clicked. The action that takes place on the button click is shown on line 5.

```
private Button mDateButton;
```

After this variable is created, you need to initialize it in the `onCreate()` method (right after the call to `setContentView()`):

```
mDateButton = (Button) findViewById(R.id.reminder_date);
```

→4   This line overrides the default click behavior of the button so that you can provide your own set of actions to perform. The `View v` parameter is the view that was clicked.

→5   This line defines what you want to happen when the button is clicked. In this instance, you call a method on the base activity class — `showDialog()`. The `showDialog()` method accepts one parameter — the ID of the dialog box that you want to show — in this case, a constant called `DATE_PICKER_DIALOG`. This is the first of two constants you must define at the top of the class file by typing the following code. (The second constant is used in the section, "Wiring up the time picker," later in this chapter.)

```
private static final int DATE_PICKER_DIALOG = 0;
private static final int TIME_PICKER_DIALOG = 1;
```

→8   This method is called to update the button text of the date and time buttons. This method is created in Listing 13-5.

→9   This method is called to update the time button text. This method is created in Listing 13-6.

Creating the showDialog() method

The `showDialog()` method performs some work for you in the base activity class, and at the end of the day, the only thing you need to know is that by calling `showDialog()` with an ID, the activity's `onCreateDialog()` method is called. At the bottom of your `ReminderEditActivity` file, type the code from Listing 13-3 to respond to the `showDialog()` method call.

Listing 13-3: Responding to showDialog() with onCreateDialog()

```
@Override
protected Dialog onCreateDialog(int id) {                         →2
    switch(id) {
        case DATE_PICKER_DIALOG:
            return showDatePicker();                             →4
    }
    return super.onCreateDialog(id);
}

private DatePickerDialog showDatePicker() {                       →10
    DatePickerDialog datePicker = new DatePickerDialog(ReminderEdit
            Activity.this, new DatePickerDialog.OnDateSetListener()
            {                                                    →13

        @Override
        public void onDateSet(DatePicker view, int year, int monthOfYear,
                int dayOfMonth) {                                →17

            mCalendar.set(Calendar.YEAR, year);                  →19
            mCalendar.set(Calendar.MONTH, monthOfYear);
            mCalendar.set(Calendar.DAY_OF_MONTH, dayOfMonth);    →21
            updateDateButtonText();                              →22
        }
    }, mCalendar.get(Calendar.YEAR), mCalendar.get(Calendar.MONTH),
            mCalendar.get(Calendar.DAY_OF_MONTH));               →25
    return datePicker;                                           →26
}

private void updateDateButtonText() {                            →29
    SimpleDateFormat dateFormat = new SimpleDateFormat(DATE_FORMAT); →30
    String dateForButton = dateFormat.format(mCalendar.getTime()); →31
    mDateButton.setText(dateForButton);                         →32
}
```

Each important line of code is explained as follows:

→2   The `onCreateDialog()` method is overridden and called when the `showDialog()` method is called with a parameter. The `int id` parameter is the ID that was passed into the `showDialog()` method previously.

→4   This line of code determines whether the ID passed into the `onCreateDialog()` is the same one that was passed in as a parameter to the `showDialog()` method. If it matches the DATE_ PICKER_DIALOG value, it returns the value of the `showDat- ePicker()` method. The `showDatePicker()` call must return a `Dialog` type for `onCreateDialog()` to show a dialog box.

→10   The `showDatePicker()` method definition that returns a `DatePickerDialog`.

→13   On this line, you create a new `DatePickerDialog` that accepts the current context as the first parameter. You provide the current instance `ReminderEditActivity.this` as the `Context`. The full class name is included because it's inside a nested statement, therefore fully qualified names are required. The next parameter is the `onDateSetListener()`, which provides a callback that is defined from line 13 through line 22. This callback provides the value of the date that was chosen through the date picker. The other parameters for the `DatePickerDialog` are listed on line 25.

→17   The implementation of the `onDateSet()` method that is called when the user sets the date through the `DatePickerDialog` and clicks the Set button. This method provides the following parameters:

- `DatePicker view`: The date picker used in the date selection dialog box

- `int year`: The year that was set

- `int monthOfYear`: The month that was set in format 0–11 for compatibility with the `Calendar` object

- `int dayOfMonth`: The day of the month

→19 **through** →21   This code block uses a variable by the name of `mCalendar`. This is a classwide `Calendar` variable that allows me to keep track of the date and time that the user set while inside the `ReminderEditActivity` through the `DatePickerDialog` and `TimePickerDialog`. You also need this variable — define a classwide `Calendar` variable at the top of the class file with the name of `mCalendar`. In this code block, you use the setter and `Calendar` constants to change the date values of the `Calendar` object to that of the values the user set through the `DatePickerDialog`.

```
private Calendar mCalendar;
mCalendar = Calendar.getInstance();
```

Inside the `onCreate()` method, provide the `mCalendar` object with a value using the `getInstance()` method. This method returns a new instance of the `Calendar` object.

→**22** After the `mCalendar` object has been updated, you make a call to `updateDateButtonText()` that updates the text of the button that the user clicked to open the `DatePickerDialog`. This method is explained on lines 29 through 31.

→**25** These are the remaining parameters to set up the `DatePickerDialog`. These calendar values are what shows when `DatePickerDialog` opens. You use the `mCalendar` get accessor to retrieve the year, month, and day values of `mCalen-dar`. If `mCalendar` has not been previously set, these values are from today's date. If `mCalendar` has previously been set and the user decides to open the `DatePickerDialog` again to change the date, the `mCalendar` object returns the values that were set from the previous date selection as the default of the new `DatePickerDialog`.

→**26** At the end of this method, you return an instance of the `Dialog` class because `onCreateDialog()` requires it. Because the `DatePickerDialog` class is a subclass of `Dialog`, you can return the `DatePickerDialog`. This allows `onCreateDialog()` to create the dialog box for the user to see onscreen.

→**29** As shown on line 22, the `updateDateButtonText()` method is called after the `mCalendar` object is set up with new date values. This method is used to update the text of the Date button that the user selects when he wants to change the date. In this method, you set the button text to the value of the date that was selected so that the user can easily see what the reminder date is without having to open the `DatePickerDialog`.

→**30** This line sets up a `SimpleDateFormat` object. This object is used to format and parse dates using a concrete class in a local-sensitive manner, such as either Gregorian or Hebrew calendars. Using the date formatting options listed in the Java documentation (`http://download-llnw.oracle.com/javase/1.4.2/docs/api/java/text/SimpleDateFormat.html`), you can provide various output. On this line, I'm using a local constant called DATE_FORMAT as a parameter to set up the `SimpleDateFormat`. This constant defines the format in which you'd like the date information to be visible to the end user. You need to define this constant at the top of the class file as follows:

```
private static final String DATE_FORMAT = "yyyy-MM-dd";
```

This date format is defined as "yyyy-MM-dd," meaning a four-digit year, a two-digit month, and a two-digit day. Each is separated by a hyphen. An example of this would be 2010-09-10.

→**31**   On this line, I use the `SimpleDateFormat` object to format the `mCalendar` date by calling the `getTime()` method on the `mCalendar` object. This method returns a date object that the `SimpleDateFormat` object parses into the `DATE_FORMAT` that you specified on line 30. You then set the result — a string result — into a local variable.

→**32**   Using the local variable you set up on line 31, you set the text of the Date button using the `Button` class's `setText()` method.

The `DatePickerDialog` widget is now wired up to accept input from the user.

Wiring up the time picker

The `TimePickerDialog` allows users to select a time during the day in which they would like to be reminded of the task at hand.

Setting up a `TimePickerDialog` is almost identical to setting up a `DatePickerDialog`, the only thing that differs is that you'll be using a different constant and showing a different dialog. The first thing you need to do is declare the `onClickListener()` for the Time button. To do so, create a local `mTimeButton` variable at the top of the class file with the following code:

```
private Button mTimeButton;
```

You then need to initialize the variable in the `onCreate()` method as follows:

```
mTimeButton = (Button) findViewById(R.id.reminder_time);
```

Setting up the Time button click listener

Now that you have a Time button to work with, you can set up the click listener for it. In the `registerButtonListenersAndSetDefaultText()` method, type the code shown in Listing 13-4.

Listing 13-4:   Implementing the Time Button's OnClickListener

```
mTimeButton.setOnClickListener(new View.OnClickListener() {
    @Override
    public void onClick(View v) {
        showDialog(TIME_PICKER_DIALOG);                          →4
    }
});
```

This entire method is the same as the Date button's `onClickListener()`, except that on line 4, you use a different constant as a parameter to the `showDialog()` method. You do this because when `showDialog()` is called, it in turn calls `onCreateDialog()` with that ID. When `showDialog()` calls `onCreateDialog()`, you can determine which dialog box to show according to the constant that you used in `showDialog()`. You need to create the `TIME_PICKER_DIALOG` constant at the top of the class file.

Now you need to go back to the `onCreateDialog()` method and add the following code after the `return showDatePicker()` code:

```
case TIME_PICKER_DIALOG:
    return showTimePicker();
```

Creating the showTimePicker() method

The `showTimePicker()` method has not been created. Create that method now by adding the full method definition in Listing 13-5.

Listing 13-5: The showTimePicker() Method

```
private TimePickerDialog showTimePicker() {
    TimePickerDialog timePicker = new TimePickerDialog(this, new
        TimePickerDialog.OnTimeSetListener() {                           →3
        @Override
        public void onTimeSet(TimePicker view, int hourOfDay, int minute)
            {                                                            →5
            mCalendar.set(Calendar.HOUR_OF_DAY, hourOfDay);             →6
            mCalendar.set(Calendar.MINUTE, minute);                    →7
            updateTimeButtonText();                                    →8
        }
    }, mCalendar.get(Calendar.HOUR_OF_DAY),                            →10
            mCalendar.get(Calendar.MINUTE), true);                     →11

    return timePicker;
}
```

The code in Listing 13-5 is fairly straightforward because it's almost identical to that of the `showDatePicker()` method. However, you can see differences on the following lines:

→3   Here a `TimePickerDialog` is being set up with a new `OnTimeSetListener()` that is called when the user sets the time with the `TimePickerDialog`.

→5   When the time is set, the hour and minute are passed into the `onTimeSet()` method, allowing you to perform necessary actions with the values.

→6   Here I am setting the classwide `Calendar` object's hour of the day.

→7   Here I am setting the classwide `Calendar` object's minute of the hour.

→8   This line delegates the updating of the Time button's text to a method called `updateTimeButtonText()`. This method is explained in Listing 13-6.

→10   This line specifies the default hour for the `TimePickerDialog`. This value is retrieved from the classwide `Calendar` object.

→11   This line specifies the default minute for the `TimePickerDialog`. This value is retrieved from the classwide `Calendar` object. The last parameter is set to the value of `true`, which informs the `TimePickerDialog` to show the time in 24-hour format as opposed to a 12-hour time format with a.m. and p.m. distinctions.

At the end of the method, the instance of the `TimePickerDialog` is returned to the `onCreateDialog()` method to allow it to show the dialog box to the end user.

On line 8, you made a call to `updateTimeButtonText()`. This method is very similar to the `updateDateButtonText()`, as shown previously in this chapter. Type the code from Listing 13-6 into the editor to create the `updateTimeButtonText()` method.

Listing 13-6: The updateTimeButtonText() Method

```
private void updateTimeButtonText() {
    SimpleDateFormat timeFormat = new SimpleDateFormat(TIME_FORMAT);    →2
    String timeForButton = timeFormat.format(mCalendar.getTime());       →3
    mTimeButton.setText(timeForButton);                                  →4
}
```

This code is explained as follows:

→2   This line of code creates a new `SimpleDateFormat`, but this time with a different constant. You need to create the `TIME_FORMAT` constant at the top of the class file as follows:

```
private static final String TIME_FORMAT = "kk:mm";
```

This constant informs the `SimpleDateFormat` class that you would like the calendar to output the minutes and seconds separated by a colon. An example would be 12:45 to represent 12:45 p.m.

→3   This line formats the current calendar's time to that of the pre-scribed format on line 2.

→4   This line updates the button text to the time that was retrieved on line 3.

At this point, you've set up the date and time picker dialog widgets to accept values from the user. The best part is, you did not have to write the date and time logic; you simply had to respond to the click listeners.

Creating Your First Alert Dialog Box

Have you ever worked with an application that did not inform you of a warning or did not alert you of something? If not, take the following example into consideration. Imagine an e-mail client that does not inform you that you have new e-mail. How annoying would that be? Alerting users of important issues or choices that need to be made is an integral part of any user experience. A few examples of where you might want to use a dialog box to inform the user of a message and/or to have the user perform an action are as follows:

- Something is happening in the background (this is what a `ProgressDialog` does).
- The values in an `EditText` view are invalid.
- The network has become unavailable.
- The user needs to select a date or time (as I just demonstrated).
- The state of the phone is not compatible with the application. Maybe the app needs to be GPS enabled or needs an SD Card, and you've detected these issues upon the application starting.
- The user needs to choose from a list of items.

Although this is not a comprehensive list, it does give you an inkling of what is possible with dialog boxes.

The Android system has a framework built around dialog boxes that enables you to provide any implementation that you may need.

Various types of dialog boxes are available, including these, the most common:

- **Alert:** Alerts the user of something important. Also allows the user to set the text value of the buttons, as well as the actions performed, when they are clicked. As a developer, you can provide the `AlertDialog` with a list of items to display — allowing the user to select from a list of items.
- **Progress:** Used to display a progress wheel or bar. This dialog box is created through the `ProgressDialog` class.
- **Custom:** A custom dialog box created and programmed by you, the master Android developer. You create a custom dialog class by extending the `Dialog` base class or through custom layout XML files.

If you ever work with any type of blocking process (network communication, long-running tasks, and so on), you should always provide the user with some type of dialog box or progress indicator letting the user know what is happening. If the user does not know something is happening, she is likely to think that the application has stopped responding and might stop using the app. The Android framework provides various progress indicators. A couple common progress classes are `ProgressDialog` and `ProgressBar`.

Although the topic is too complex to cover here, the `AsyncTask` class is the class that you would use to help manage long-running tasks while updating the user interface. A great tutorial for this class exists in the Android documentation under Painless Threading, located at `http://d.android.com/resources/articles/painless-threading.html`. You can also create a new thread in code, but the `AsyncTask` class helps simplify this process.

Choosing the right dialog box for a task

It's up to you to determine which dialog box you should use for each given scenario, but I follow a logical series of steps to determine which dialog box to use. (See Figure 13-2.)

Creating your own alert dialog box

At times, you need to notify the user of something important, and to do so, you need to present them with a dialog box. Android has made this very simple with the introduction of the `AlertDialog.Builder` class. This class allows you to easily create an `AlertDialog` with various options and buttons. You can react to these button clicks through the `onClickListener()` of each button.

The `AlertDialog.Builder` class is not used in the Task Reminder application. However, I demonstrate how to create one in Listing 13-7.

Assume that the user clicked the Save button on the Task Reminder application and that you wanted to pop up a confirmation window that resembles Figure 13-3, asking the user whether he is sure that he wants to save.

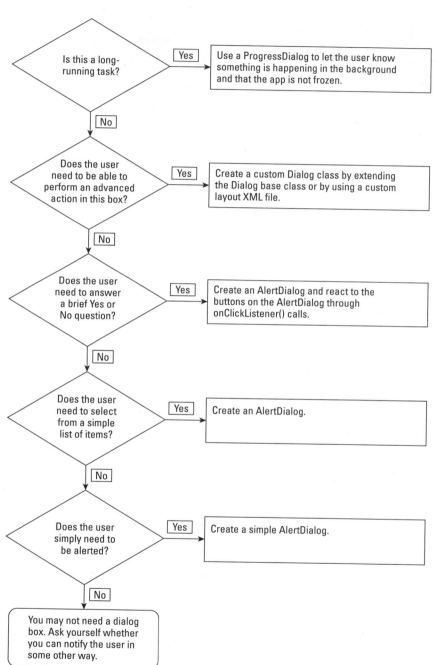

Figure 13-2:
An EditText
view
without the
gravity set.

Figure 13-3:
The
confirmation
Alert
Dialog
window.

To present an `AlertDialog` in this manner, you need to set a click listener for the Save button. Inside that click listener, you create the dialog box as shown in Listing 13-7.

Showing dialog boxes through the `showDialog()` and `onCreateDialog()` mechanisms is best. However, for brevity, here you create the dialog box inside the Save button click listener.

Listing 13-7: Creating an AlertDialog with the AlertDialog.Builder Class

```
AlertDialog.Builder builder
    = new AlertDialog.Builder(ReminderEditActivity.this);          →2
builder.setMessage("Are you sure you want to save the task?")      →3
        .setTitle("Are you sure?")                                 →4
        .setCancelable(false)                                      →5
        .setPositiveButton("Yes",                                  →6
        new DialogInterface.OnClickListener() {                    →7
            public void onClick(DialogInterface dialog, int id) {
                    // Perform some action such as saving the item →9
            }
        })
        .setNegativeButton(„No", new DialogInterface.OnClickListener() {  →12
            public void onClick(DialogInterface dialog, int id) {
                        dialog.cancel();                           →14
            }
        }
});
builder.create().show();                                           →17
```

This code is explained as follows:

→**2**   This line sets up the `AlertDialog.Builder` class with the context of the `AlertDialog.Builder` as the current running activity. In this case, it's `ReminderEditActivity`.

→**3**   This line sets the message that will show in the middle of the `AlertDialog` (as shown in Figure 13-2). This value can be a string or a string resource.

→**4**   This line sets the title of the `AlertDialog`. This value can be a string or a string resource.

→**5**   This line sets the cancelable attribute to false. This means that the user is required to make a selection by using the buttons on the `AlertDialog`. The user cannot click the Back button on the device to exit the `AlertDialog` if this flag is set to false.

→**6**   This line sets the positive button text. The positive button is the button that the user clicks when she wants to perform the action as indicated in the `AlertDialog`. In this case, it is set to Yes, indicating that the user wants to perform the action. This value can be a string or a string resource.

→**7**   This block of code that starts on line 7 and ends on line 11 is the definition of the `onClickListener()` for the positive (Yes) button. When the button is clicked, this code executes. A comment is included on line 9 indicating where your code would go.

→**12**   This line sets the negative button text. The negative button is the button that indicates that the user does not want to perform the action that is being requested through the `AlertDialog`. I have set the text value of this button to No. This value can be a string or a string resource.

→**14**   This line is the `onClickListener()` for the negative button. The listener provides a reference to the dialog box that is currently shown. I am calling the `cancel()` method on the `Dialog` object to close the dialog box when the user clicks No on the `AlertDialog`.

→**17**   This line informs Android to create the `Dialog` through the `create()` method and then informs Android to show the dialog box to the end user with the `show()` method. This projects the `AlertDialog` onto the screen.

Creating a dialog box with the `AlertDialog.Builder` class makes life a lot easier than having to derive your own `Dialog` class. If at all possible, create your dialog box with the `AlertDialog.Builder` class because it gives your application a consistent user experience that is familiar to Android users.

When the user clicks the Save button (or whatever button you've attached this code to), he receives an `AlertDialog` confirming that he wants to save the task. I am not saving the task in this instance, but code could be provided to save the task as demonstrated in Chapter 14, when I save the task to an SQLite database.

Other options also exist on the `Dialog` class and can be found with great examples here: `http://d.android.com/guide/topics/ui/dialogs.html`.

Validating Input

You've created your form so that users can enter information, and perhaps you've already created the mechanism to save the content to a database or remote server. But what happens when the user enters invalid text or no text? This is where input validation enters the picture.

Input validation validates the input before the save takes place. Assume that the user does not enter text for the title or the message and attempts to save; should she be allowed to save? Of course not!

Unfortunately, a built-in Android validation framework does not exist. Hopefully, in future versions of the Android platform this feature will be introduced. However, you have ways to validate input with the current framework.

The method in which you provide validation to the user is up to you. Here are some common methods in which I've seen developers implement validation:

✔ `TextWatcher`: Implement a `TextWatcher` on the `EditText` widget. This class provides callbacks to you each time the text changes in the `EditText` widget. Therefore, you can inspect the text on each keystroke.

✔ `On Save`: When the user attempts to save the form that he is working with, inspect all the form fields at that time and inform the user of any issues found.

✔ `onFocusChanged()`: Inspect the values of the form when the `onFocusChanged()` event is called — which is called when the view has focus and when it loses focus. This is usually a good place to set up validation.

The Task Reminder application does not provide input validation; however, you can add validation via one of the methods described previously.

Toasting the user

The most common way to inform the user that something is incorrect is to provide a Toast message to her. A Toast message pops onto the screen for a short period of time informing the user of some type of information — in this case, an error with input values.

Providing a Toast is as simple as implementing the following code, where you'd like to inform the user of the input error:

```
Toast.makeText(ReminderEditActivity.this, "Title must be filled in", Toast.
           LENGTH_SHORT).show();
```

You might show this Toast message when the user does not enter a title into the title field and when the user clicks the Save button.

The only issue with Toast messages is that they are short-lived by default, yet they can be configured to display longer. If the user happens to glance away for a moment, he can miss the message because the Toast message only shows up for a few moments and then fades out.

Using other validation techniques

A Toast message is not the only way to inform users of a problem with their input. A couple of other popular validation techniques are as follows:

- ✔ AlertDialog: Create an instance of an AlertDialog that informs the user of the errors. This method ensures that the user will see the error message because the alert must either be canceled or accepted.

- ✔ **Input-field highlighting:** If the field is invalid, the input field (the EditText widget) could have its background color changed to red (or any color you choose) to indicate that the value is incorrect.

- ✔ **Custom validation:** If you're feeling adventurous, you could create a custom validation library that would handle validations of all sorts. This validation could highlight the field and draw small views with arrows pointing to the error with highlighting. Google uses similar validation for its sign-in window when you log on to a device (such as the Samsung Galaxy S) for the first time.

I've shown the most common methods of displaying input validation information. But as long as you can dream up new ways to inform users of an error, you can use those new methods. For example, in Chapter 15, I introduce you to the notification bar. I have used the notification bar in my own projects to inform users of a problem with a background service. Although this is a special case, it is a valid one, and it provides the user with feedback that he needs to make adjustments to the application or the workflow.

Chapter 14

Getting Persistent with Data Storage

Most applications these days require you to save information for later use. The Task Reminder application would not be that useful if it did not save the tasks, now would it? Thankfully, the Android platform — in combination with Java — provides a robust set of tools that you can use to store data.

This chapter delves deeply into creating and updating an SQLite database. Although in this chapter I explain the steps necessary to do these things, I do not provide a lot of database theory. If you're not familiar with the SQL language or the SQL database, I advise you to review the SQLite website for more information: `www.sqlite.org`.

This chapter is also very code intensive, and if you find yourself getting lost, feel free to reference the completed application source code — available online at this book's companion website.

Finding Places to Put Data

Depending on the requirements of your application, you may need to store your data in a variety of places. For example, if your application includes music files, users may want to play those files with other music programs — therefore, you'd want to store the files in a location where other applications can access them such as the SD Card (which is considered world readable and writeable). On the other hand, your application may need to store sensitive data such as encrypted username and password details. For these types of applications, you wouldn't share data — you would place it in a secure local storage environment. Regardless of your situation, Android provides various locations where you can persist your data. The most common are as follows:

- ✔ **Shared preferences:** Shared preferences are private data stored in key-value pairs that are available to all components of your application — and that application only. I cover how to use preferences in Chapter 17.

- ✔ **Internal storage:** You can save files on the device's internal storage. By default, files stored in internal storage are private to your application, and other applications cannot access them (neither can the user of the device). When the user uninstalls the application, the private files are removed. You might want to use internal storage as a workspace for when you need to process files at different times during usage of your app over a long period of time.

- ✔ **Local cache:** If you want to cache some data rather than store it persistently, the internal data directory is where you should create the cache. You should use the `getCacheDir()` method (which is available on the `Activity` or `Context` objects in Android), which provides you with an application-specific cache directory for you to store files. Note that if you store data here and the system gets low on internal storage space, Android may delete these files to reclaim space. To prevent this from happening, try to stay within a reasonable limit of space consumed of around 1 MB.

- ✔ **External storage:** Every Android device supports shared external storage that you can use to store files. This storage can either be removable storage, such as a Secure Digital Card (SD Card), or nonremovable internal storage. Files saved to external storage are public — meaning that anyone or any application can alter these files. No security is enforced upon external files. The files can be modified by the user either through a file-manager application or by connecting the user's device to a computer through a USB cable and mounting the device as external storage. Before you work with external storage, check the current state of the external storage with the `Environment` object, using a call to `getExternalStorageState()` to check whether the media is available. You would store files such as music, wallpapers, and photos in external storage.

  In Android 2.2, a new set of methods was introduced to handle external files. The main method is a call on the `Context` object — `getExternalFilesDir()`. This call takes a string parameter as a key to help define what kind of media you are going to save, such as ringtones, music, photos, and so on. For more information, view the external data storage examples and documents here: `http://d.android.com/guide/topics/data/data-storage.html#filesExternal`.

✔ **SQLite database:** Android supports the full use of SQLite databases. An SQLite database is a lightweight SQL (Structured Query Language) database implementation that is available across various platforms including Android, iPhone, Windows, Linux, Mac, and various other embedded devices. You can create tables and perform SQL queries against the tables accordingly. You implement an SQLite database in this chapter to handle the persistence of the tasks in the Task Reminder application.

✔ **Network connection:** Last but definitely not least is network storage, also known as *remote storage.* This type of storage can be any remote data source that you can access. For example, Flickr exposes an API that allows you to store images on its servers. Your application could work with Flickr to store your images. Your application may also be an Android application for a popular tool on the Internet, such as Twitter, Facebook, or Basecamp. Your app would communicate through HTTP (or any other protocol you deem necessary) to send information to the third-party APIs to store the data.

These locations offer up quite the palette of data storage options. However, you need to figure out which one you want to use. At times, you want to use multiple storage mechanisms.

For example, if you have an application that communicates with a third-party remote API, such as Twitter, you may want to keep a local copy of all the data since your last update with the server because network communication is slow and is not 100 percent reliable. Doing this allows the application to remain usable (in some fashion) until you can make the next update. You can store the data in a local copy of an SQLite database, and then when the user initiates an update, the new updates would refresh the SQLite database with the new data.

If your application relies solely on network communication for retrieval and storage of information, you may want to consider using the SQLite database (or any other storage mechanism) to keep the application usable when the user is not able to connect to a network (most developers know this as *offline-mode*). You'd be surprised how often this happens. If your application doesn't function when a network connection is unavailable, you will most likely receive negative reviews in the Android Market (as well as a lot of feature requests to make it work offline). Although this does introduce quite a bit of extra work into your application development process, it's worth it tenfold in user experience.

Asking the User for Permission

You wouldn't want your next-door neighbor storing his holiday decorations in your storage shed without clearing it through you first, would you? I didn't think so! Android is no different — storing data anywhere on the device requires some sort of permission from the user. But that's not the only thing that requires some sort of permission.

Seeing how permissions affect the user experience

When users install applications from the Android Market, the application's manifest file is inspected for required permissions that the application needs to operate. Anytime your application needs access to sensitive components, such as external storage, access to the Internet, phone device information, and so on, the user is notified that the application would like to access these components. The user decides whether she would like to install the application.

If your application requests a lot of unnecessary permissions, the user may get suspicious and choose not to install the application. Imagine if the Screen Brightness Toggle application (built previously in this book) was requesting your current GPS location, needed access to the Internet, and wanted to know information about the device (such as hardware info). The Screen Brightness Toggle application has no need for those permissions. Users tend to be wary of installing an application that is overzealously requesting permissions.

Through the many different applications that I've published, I've found that the fewer number of permissions your application requests, the more likely the user is to install your application. If your application does not need the permission, yank the permission out of the application.

Setting requested permissions in the AndroidManifest.xml file

When you need to request permissions, you need to add them to the `AndroidManifest.xml` file in your project. No permission is necessary to work with an SQLite database; therefore, I'm going to add two permissions to the Task Reminder application that will be required when I add the alarm manager code in Chapter 15:

- ✔ `android.permission.RECEIVE_BOOT_COMPLETED`: This permission allows the application access to know when the tablet reboots.
- ✔ `android.permission.WAKE_LOCK`: This permission keeps the tablet "awake" while it performs some background processing.

These items are covered in detail, along with the `AlarmManager`, in Chapter 15.

The RECEIVE_BOOT_COMPLETED and WAKE_LOCK permissions mentioned above are unique and are used infrequently — so here I outline a couple of the most common permissions and describe how to set them. A lot of applications require access to the Internet to operate. Some applications also need to write data to the SD Card. If you need either of these, add the following permissions:

✔ **Internet:** `android.permission.INTERNET`

✔ **SD Card:** `android.permission.WRITE_EXTERNAL_STORAGE`

You can add permissions to the `AndroidManifest.xml` file in one of two ways:

✔ **Through the `AndroidManifest.xml` Permissions Editor.** Choose Add➪Uses Permission and then choose the permission from the drop-down list.

✔ **Through manually editing the XML file.** This method is how I prefer to do it. You need to add a `uses-permission` element to the manifest `element`. The XML permission request looks like this:

```
<uses-permission android:name="android.permission.WAKE_LOCK" />
```

If you have not done so already, add the WAKE_LOCK and RECEIVE_BOOT_COMPLETED permissions to the Task Reminder application. To view a full list of available permissions, view the Android permission documentation here: `http://d.android.com/reference/android/Manifest.permission.html`.

If you do not declare the permissions your application needs and a user installs your application on his device, your application will not function as expected; sometimes run-time exceptions will be thrown and crash your application. Always be sure to check that your permissions are present. Most likely, they will be because the app will not work on a device or emulator if they are not.

Creating Your Application's SQLite Database

The Task Reminder application needs a place to store and retrieve the user's tasks, and the best place for this kind of information is inside an SQLite database. Your application needs to read, create, update, and delete tasks from the database.

The Create, Read, Update, and Delete actions are known as CRUD operations — each letter standing for its respective action.

Understanding how the SQLite database works

The two activities in the Task Reminder application need to perform various duties on the database to operate. `ReminderEditActivity` needs to do the following:

- ✔ Create a new record.
- ✔ Read a record so that it can display the details for editing.
- ✔ Update the existing record.

The `ReminderListActivity` needs to perform these duties:

- ✔ Read all the tasks to display them on the screen.
- ✔ Delete a task by responding to the click event from the context menu after a user has long-pressed an item.

To work with an SQLite database, you must communicate with SQLite through classes in the `android.database` package. Because common practice is to abstract as much of the database communication away from the `Activity` objects as possible, the database mechanisms are placed into another Java file (and usually a package if the database portion is quite large) to help separate the application into layers of functionality. Therefore, if you ever need to alter code that affects the database, you know that you need to change the code in only one location. You follow this approach in the sections that follow.

Creating a Java file to hold the database code

Your first task is to create a Java file in your Android project that will house all the database-centric code. I named my file `RemindersDbAdapter.java` — I chose this name because this is a simple implementation of the *adapter* software engineering pattern.

The adapter pattern is simply a wrapper class that allows incompatible classes to communicate with each other. Think of the adapter pattern as the wires and ports behind your television and DVD player. The cables plug into ports that are essentially adapters that allow devices to communicate that normally couldn't. They share a common interface. By creating an adapter to handle the database communication, you can program with Java. At run time, the adapter class does the translation and adapts certain Java requests into SQLite-specific commands.

Defining the key elements

Before you open and create your database, I need to define a few key fields. Type the code from Listing 14-1 into your `RemindersDbAdapter` class.

Listing 14-1:   The Constants, Fields, and Constructors of the RemindersDbAdapter Class

```
private static final String DATABASE_NAME = "data";                    →1
private static final String DATABASE_TABLE = "reminders";              →2
private static final int DATABASE_VERSION = 1;                          →3

public static final String KEY_TITLE = "title";                        →5
public static final String KEY_BODY = "body";
public static final String KEY_DATE_TIME = "reminder_date_time";
public static final String KEY_ROWID = "_id";                          →8


private DatabaseHelper mDbHelper;                                       →11
private SQLiteDatabase mDb;                                             →12


private static final String DATABASE_CREATE =                          →14
        "create table " + DATABASE_TABLE + " ("
                + KEY_ROWID + " integer primary key autoincrement, "
                + KEY_TITLE + " text not null, "
                + KEY_BODY + " text not null, "
                + KEY_DATE_TIME + " text not null);";

private final Context mCtx;                                            →21

public RemindersDbAdapter(Context ctx) {                              →23
    this.mCtx = ctx;
}
```

Each line is explained in detail here:

→1   This line is the physical name of the database that will exist in the Android file system.

→2   This line is the name of the database table that will hold the tasks. For more on how to set up this table, see the section, "Visualizing the SQL table," later in this chapter.

→3   This line is the version of the database. If you were to update the schema of your database, you would increment this and provide an implementation of the onUpgrade() method of the DatabaseHelper. You create the database helper in the "Creating the database table," section later in this chapter.

→5-→8   These lines define the column names of the table that is described in the "Visualizing the SQL table" section that follows.

→11   This line is the classwide DatabaseHelper instance variable. The DatabaseHelper is an implementation of the SQLiteOpenHelper class in Android. The SQLiteOpenHelper class helps with the creation and version management of the SQLite database; the DatabaseHelper implementation will be covered later in this chapter.

→12   This line is the class-level instance of the SQLite database object that allows you to create, read, update, and delete records.

→14   This line defines the create script for the database. I'm concatenating the various values from the previous lines to create the various columns. Each component of the script is explained in the "Visualizing the SQL table," section that follows.

→21   This line is the Context object that will be associated with the SQLite database object.

→23   The Context object is set through the constructor of the class.

The SQL database is now ready to be created with the DATABASE_CREATE script as defined previously.

Visualizing the SQL table

The table object in SQL is the construct that holds the data that you decide to manage. Visualize tables in SQLite as similar to spreadsheets. Each row consists of data, and each column represents the data inside the row. In Listing 14-1 on lines 5–8, you defined the column names for the database. These column names equate to the header values in a spreadsheet, as shown in Figure 14-1. Each row contains a value for each column, which is how your data is stored in SQLite (conceptually, anyway — the real data is stored as 1s and 0s).

Figure 14-1:
Visualizing
data in
the Task
Reminder
application.

_id	title	body	reminder_date_time
1	Order Flight Tickets	Go to travel site and order ti...	2010-11-15 16:15
2	Schedule Time Off	Email manager at work to ...	2010-11-17 15:00
3	Take Vacation	YES! Finally take that much n...	2010-12-10 14:30
4	Pay Bills	Pay the bills through bill pay.	2010-11-10 12:15

Starting on line 14, you assemble the database create script. This script concatenates various constants from within the file to create a database create script. When you run this script in SQLite, SQLite creates a table by the name of `reminders` in a database called `data`. The columns and how they're built in the create script are described as follows:

✔ `create table DATABASE_TABLE`: This portion of the script informs SQLite that you want to create a database table with the name of `reminders`.

✔ `ROW_ID`: This property acts as the identifier for the task. This column has the `integer primary key autoincrement` attributes applied to it. The `integer` attribute specifies that the row is an integer. The `primary key` attribute states that the `ROW_ID` is the primary identifier for a task. The `autoincrement` attribute informs SQLite that each time a new task is inserted, simply automatically increment the `ROW_ID` to the next available integer. For example, if rows 1, 2, and 3 existed and you inserted another record, the value of the `ROW_ID` column in the next row would be 4.

✔ `KEY_TITLE`: This column is the title of the task that the user provides, such as "Schedule Vacation." The text attribute informs SQLite that you are working with a text column. The `not null` attribute states that the value of this column cannot be null — you must provide a value.

✔ `KEY_BODY`: This portion is the body or description of the task. The attributes for this column are the same as for `KEY_TITLE`.

✔ `KEY_DATE_TIME`: This column is where the date and time of the reminder are stored. The attributes are the same as the prior two columns. Wait! You're probably thinking, "This is a date field; why is he storing it as text?" This is because SQLite does not have a storage class associated with storing dates or times.

For more information on dates and times in SQLite, view the documentation at `www.sqlite.org/datatype3.html#datetime`.

Creating the database table

You're now ready to create your first table. To do so, you provide an implementation of SQLiteOpenHelper. In the RemindersDbAdapter class, type the code as shown in Listing 14-2. This listing creates a nested Java class inside the RemindersDbAdapter class. You can type this at the end of the class.

Listing 14-2:   Creating Your First Database Table

```
private static class DatabaseHelper extends SQLiteOpenHelper {       →1
       DatabaseHelper(Context context) {
              super(context, DATABASE_NAME, null, DATABASE_VERSION);    →3
       }

       @Override
       public void onCreate(SQLiteDatabase db) {                        →7
              db.execSQL(DATABASE_CREATE);                              →8
       }

       @Override
       public void onUpgrade(SQLiteDatabase db, int oldVersion,
                       int newVersion) {                                →12
           // Not used, but you could upgrade the database with ALTER
           // Scripts
       }
}
```

The code lines are explained here:

→1   This line is the implementation of your SQLiteOpenHelper.

→3   The call made to the base SQLiteOpenHelper constructor. This call creates, opens, or manages a database. The database is not actually created or opened until getReadableDatabase() or getWriteableDatabase() is called on the SQLiteOpenHelper instance — in this case, it would be the mDbHelper variable.

→7   The onCreate() method, which is called when the database is created for the first time.

→8   This line is where all the magic happens — it creates your database and your database table. The execSQL() method accepts an SQL script string as a parameter. This is the SQL that the SQLite database executes to create the database table.

→12   The onUpgrade() method is used when you need to upgrade an existing database.

You now create the database by calling the `getReadableDatabase()` or `getWritableDatabase()` method on the `DatabaseHelper` object. To do this, type the following code anywhere into your `RemindersDbAdapter` class:

```
public RemindersDbAdapter open() throws android.database.SQLException {
        mDbHelper = new DatabaseHelper(mCtx);
        mDb = mDbHelper.getWritableDatabase();
        return this;
}
```

The `open()` method opens (and creates if necessary) the database using the `DatabaseHelper()` class that was just created. This class then returns itself through the `this` Java keyword. The reason that the class is returning itself is because the caller (`ReminderEditActivity` or `ReminderListActivity`) needs to access data from this class and this method returns an instance of the `RemindersDbAdapter`.

Closing the database

A database is an expensive resource and should be closed when not in use. To close the database, type the following method anywhere into your `RemindersDbAdapter` class:

```
public void close() {
        mDbHelper.close();
}
```

This method closes the database when called. You call this from within the `ReminderEditActivity` when the user cancels the activity by using the Back button on the device.

Upgrading your database

When would you upgrade your database? Consider the following situation: You have released your application and 10,000 users have installed it and are actively using it. Not only are they using it, but they also love it! Some users are even sending in feature requests. You decide that you want to implement one of these feature requests, but to do so, you must change the database schema. To update your database, you must perform SQL `ALTER` statements inside the `onUpgrade()` call. Some developers upgrade their databases by dropping the existing database and then creating a new one. However, don't do this — it deletes all the user's data! Imagine updating your favorite Task Reminder application and then finding out that the upgrade erased all your preexisting tasks! That would be a major bug.

Creating and Editing Tasks with SQLite

The first thing you need to do is create a task. To do that, you need to insert a record. After that, you need to list the task(s) on the `ReminderListActivity`, which in turn allows you to tap a task to edit it, or long-press a task to delete it. These user interactions describe the create, read, update, and delete operations for your Task Reminder application. The next few sections show you how to add this functionality.

Inserting your first task entry

Inserting tasks into the SQLite database is a fairly easy process after you get the hang of it. In general terms, you simply follow these steps:

1. **Set up the required local variables.**

2. **Build the Save button click listener.**

3. **Retrieve values from** `EditText` **views.**

4. **Interact with the** `RemindersDbAdapter` **class.**

5. **Open and close the database.**

By learning how to insert your first task, you should get a good enough grasp on the `SQLiteDatabase` class interaction to perform the other necessary tasks. Therefore, in this section I introduce you to the entire implementation of `RemindersDbAdapter`, which outlines all the CRUD operations. Afterwards, I describe the CRUD tasks in greater detail.

Saving values from the screen to the database

When the user creates a task, it takes place in the `ReminderEditActivity`. To create the task, you need to create a class-level `RemindersDbAdapter` variable that is instantiated in the `onCreate()` method. After it is instantiated, you open the database with a call to the `RemindersDbAdapter`'s `open()` method in the `onResume()` method. (Type this code into your `RemindersEditActivity` class now.)

```
@Override
protected void onCreate(Bundle savedInstanceState) {
        super.onCreate(savedInstanceState);

        mDbHelper = new RemindersDbAdapter(this);
        mDbHelper.open();

        setContentView(R.layout.reminder_edit);
        // ... the remainder of the onCreate() method
```

At this point, you have a reference to the `RemindersDbAdapter` that allows you to call into the `RemindersDbAdapter` class to create a task. To add the task, you need the title, description, and reminder date and time. To gain access to the title and description, you need to add three class-level variables to `ReminderEditActivity`. Two of them are `EditText` type variables that reference the `EditText` values in the layout for the `ReminderEditActivity`. The remaining class-level variable is the Save button that you will click when you are ready to save the task to the SQLite database. I have included these variables at the top of my `ReminderEditActivity` file like this. Include these declarations in your file as well:

```
private EditText mTitleText;
private Button mConfirmButton;
private EditText mBodyText;
```

You need to instantiate those in the `onCreate()` method call like this:

```
mConfirmButton = (Button) findViewById(R.id.confirm);
mTitleText = (EditText) findViewById(R.id.title);
mBodyText = (EditText) findViewById(R.id.body);
```

You already have a `Calendar` object that was populated from the `DatePicker` and `TimePicker`; therefore, you do not need to create anything in regard to those values. The only thing left is to provide the ability to save the task after you type values into the `EditText` fields (title and description) by pressing the Save button on the screen. To do that, you need to attach a click listener to the Save button by typing the following code into the `registerButtonListenersAndSetDefaultText()` method:

```
mConfirmButton.setOnClickListener(new View.OnClickListener() {
        public void onClick(View view) {
            saveState();                                              →3
            setResult(RESULT_OK);                                    →4
            Toast.makeText(ReminderEditActivity.this,                →5
            getString(R.string.task_saved_message),
              Toast.LENGTH_SHORT).show();
            finish();                                                →8
        }
});
```

This code is explained as follows:

→3   This line of code calls the `saveState()` method.

→4   This line sets the result of the `ReminderEditActivity`. Remember, the `ReminderEditActivity` started from the call from `startActivityForResult()` inside the `ReminderListActivity`. Setting a result of `RESULT_OK` within `setResult()` informs the `ReminderListActivity` that everything went as planned when the `ReminderEditActivity` finish() method runs on line 8. The RESULT_OK constant is a member of the parent Activity class. This result code can be inspected on ReminderListActivity in the onActivity-Result() method. Your application can return any number of results to the caller to help you logically figure out what to do in the application next.

→5   This line creates a Toast message to the user, letting him know that the task saved. You need to create a string resource by the name of task_saved_message. I have chosen the value of the resource to be "Task saved."

→8   This line of code calls the finish() method, which closes the ReminderEditActivity.

You need to create the saveState() method in the ReminderEditActivity, as shown in Listing 14-3. This method communicates with the RemindersDbAdapter to save the task.

Listing 14-3: The saveState() Method

```
private void saveState() {
        String title = mTitleText.getText().toString();          →2
        String body = mBodyText.getText().toString();            →3

        SimpleDateFormat dateTimeFormat = new
            SimpleDateFormat(DATE_TIME_FORMAT);                  →6
        String reminderDateTime =
                dateTimeFormat.format(mCalendar.getTime());     →8

        long id = mDbHelper.createReminder(title, body, reminderDateTime);  →10

}
```

The lines of code are explained as follows:

→2–→3   These lines of code retrieve the text from the EditText views.

→6   This line of text defines a SimpleDateFormat that you will use to store the date and time inside the SQLite database. The DATE_TIME_FORMAT constant is used. You need to create this at the top of your class file. The code for the constant is as follows:

```
public static final String DATE_TIME_FORMAT = "yyyy-MM-dd kk:mm:ss";
```

This code defines a date and time format that could be demonstrated as 2010-11-02 12:34:21. This is a good way to store the date and time in an SQLite database.

→8   This line gets the date and time and places them into a local variable.

→**10**   This line of code creates a reminder through the `createRe-eminder()` method on the `ReminderDbAdapter` class-level variable — `mDbHelper`. You need to create that method in the `RemindersDbAdapter` class, as shown on line 38 in Listing 14-4 (see the next section).

The task is created by taking the values from the `EditText` fields and the local `Calendar` object and calling the `createReminder()` on the `RemindersDbAdapter` class. Following the adapter pattern has allowed you to wrap the SQLite logic behind a Java class, which allows the `ReminderEditActivity` to have no knowledge of the inner workings of the SQLite database.

The entire RemindersDbAdapter implementation

Have you ever bought a car after only seeing pictures of the door handle, hood, and then maybe a seat? Probably not! You'd probably never buy a car from someone who never showed you pictures of the whole thing first. Heck, you probably wouldn't even go look at it! Sometimes it's best to see everything all at once instead of in piecemeal form. Working with SQLite in the `RemindersDbAdapter` class is no different.

Trying to explain everything to you piece by piece first would not make a lot of sense; therefore, I'm going to show you the entire implementation of the `RemindersDbAdapter` in Listing 14-4 so that you can get a feel for what you're working with. Then, I explain each new area, and I cross-reference it throughout the rest of this chapter. Hopefully, this explanation helps everything gel inside that Android brain of yours.

Listing 14-4:   The Full Implementation of RemindersDbAdapter

```
public class RemindersDbAdapter {

    private static final String DATABASE_NAME = "data";
    private static final String DATABASE_TABLE = "reminders";
    private static final int DATABASE_VERSION = 1;

    public static final String KEY_TITLE = "title";
    public static final String KEY_BODY = "body";
    public static final String KEY_DATE_TIME = "reminder_date_time";
    public static final String KEY_ROWID = "_id";

    private DatabaseHelper mDbHelper;
    private SQLiteDatabase mDb;

    private static final String DATABASE_CREATE =
            "create table " + DATABASE_TABLE + " ("
                    + KEY_ROWID + " integer primary key autoincrement, "
                    + KEY_TITLE + " text not null, "
```

(continued)

Listing 14-4 *(continued)*

```
                    + KEY_BODY + " text not null, "
                    + KEY_DATE_TIME + " text not null);";

    private final Context mCtx;

    public RemindersDbAdapter(Context ctx) {
        this.mCtx = ctx;
    }

    public RemindersDbAdapter open() throws android.database.SQLException {
        mDbHelper = new DatabaseHelper(mCtx);
        mDb = mDbHelper.getWritableDatabase();
            return this;
    }

public void close() {
        mDbHelper.close();
    }

    public long createReminder(String title, String body, String
            reminderDateTime) {                                          →38
        ContentValues initialValues = new ContentValues();
        initialValues.put(KEY_TITLE, title);
        initialValues.put(KEY_BODY, body);
        initialValues.put(KEY_DATE_TIME, reminderDateTime);

        return mDb.insert(DATABASE_TABLE, null, initialValues);         →44
    }

    public boolean deleteReminder(long rowId) {                         →47
      return
        mDb.delete(DATABASE_TABLE, KEY_ROWID + "=" + rowId, null) > 0;   →49
    }

    public Cursor fetchAllReminders() {                                 →52
        return mDb.query(DATABASE_TABLE, new String[] {KEY_ROWID, KEY_TITLE,
                KEY_BODY, KEY_DATE_TIME}, null, null, null, null, null);
    }

    public Cursor fetchReminder(long rowId) throws SQLException {       →57
        Cursor mCursor =
                mDb.query(true, DATABASE_TABLE, new String[] {KEY_ROWID,
                    KEY_TITLE, KEY_BODY, KEY_DATE_TIME}, KEY_ROWID + "=" +
                rowId, null,
                        null, null, null, null);                        →61
        if (mCursor != null) {
            mCursor.moveToFirst();                                      →63
        }
```

```
        return mCursor;

    }

    public boolean updateReminder(long rowId, String title, String body, String
            reminderDateTime) {                                          →69
        ContentValues args = new ContentValues();                       →70
        args.put(KEY_TITLE, title);
        args.put(KEY_BODY, body);
        args.put(KEY_DATE_TIME, reminderDateTime);

        return
        mDb.update(DATABASE_TABLE, args, KEY_ROWID + "=" + rowId, null) > 0;  →76
    }

            // The SQLiteOpenHelper class was omitted for brevity
            // That code goes here.
}
```

→**38**   On this line, the `createReminder()` method is created. Directly below the declaration, the `ContentValues` object is used to define the values for the various columns in the database row that you will be inserting.

→**44**   On this line, the call to `insert()` is made to insert the row into the database. This method returns a long, which is the unique identifier of the row that was just inserted into the database. In the `ReminderEditActivity`, this is set to a local variable that is used in Chapter 15 to help the `AlarmManager` class figure out which task it's working with. The use of the `insert` method and its parameters are explained in detail in the following section.

→**47**   Here, the `deleteReminder()` method is defined — this method accepts one parameter, the `rowId` of the task to delete.

→**49**   Using the `rowId`, you make a call to the `delete()` method on the SQLite database to delete a task from the database. The usage and parameters of the `delete()` method are described in detail in the "Understanding the delete operation" section, later in this chapter.

→**52**   On this line, you define the `fetchAllReminders()` method, which uses the `query()` method on the SQLite database to find all the reminders in the system. The `Cursor` object is used by the calling application to retrieve values from the result set that was returned from the `query()` method call. The `query()` method usage and its parameters are explained in detail in the "Understanding the query (read) operation" section, later in this chapter.

→**57**   On this line, you define the `fetchReminder()` method, which
           accepts one parameter — the `row Id` of the task in the database
           to fetch.

→**61**   This line uses the SQLite `query()` method to return a `Cursor`
           object. The `query()` method usage and its parameters are
           explained in detail in the "Understanding the query (read) opera-
           tion" section, later in this chapter.

→**63**   The `Cursor` object can contain many rows; however, the initial
           position is not on the first record. The `moveToFirst()` method
           on the cursor instructs the cursor to go to the first record in the
           result set. This method is called only if the cursor is not null.
           The reason the cursor is not immediately positioned on the first
           record is because it's a result set. Before you can work with the
           record, you must navigate to it. Think of the result set like a box of
           items: You can't work with an item until you take it out of the box.

→**69**   On this line, you define the `updateReminder()` method that uses
           the `update()` method. The `update()` method is responsible for
           updating an existing task with new information. The `update()`
           method usage and parameters are explained in detail in the
           "Understanding the update operation" section, later in this chapter.

→**70**   The `ContentValues` object is created. This object stores the
           various values that need to get updated in the SQLite database.

→**76**   This line updates the database record with new values that were
           provided by the end user of the application. The `update()`
           method usage and its parameters are explained in detail in the
           "Understanding the update operation" section, later in this chapter.

This code listing outlines the various CRUD routines. Each accepts a variety
of different parameters that are explained in detail in the following sections.

Understanding the insert operation

The insert operation is a simple operation because you are just inserting a
value into the database. The `insert()` method accepts the following parameters:

- `table`: This parameter is the name of the table to insert the data into.
  I'm using the `DATABASE_TABLE` constant for the value.

- `nullColumnHack`: SQL does not allow inserting a completely empty
  row, so if the `ContentValues` parameter (next parameter) is empty,
  this column is explicitly assigned a `NULL` value. You pass in null for
  this value.

- `values`: This parameter defines the initial values as defined as a
  `ContentValues` object. You provide the `initialValues` local variable as the value for this parameter. This variable contains the key-value
  pair information for defining a new row.

Understanding the query (read) operation

The query operation is also known as the read operation because most of the time, you will be *reading* data from the database with the `query()` method. The query method is responsible for providing a result set based upon a list of criteria that you provide. This method returns a `Cursor` that provides random read-write access to the result set returned by the query. The query method accepts the following parameters:

- `distinct`: You want each row to be unique, so you provide `true` for this value.

- `table`: The name of the database table to perform the query against. The value you provide comes from the `DATABASE_TABLE` constant.

- `columns`: A list of columns to return from the query. Passing null returns all columns, which is normally discouraged to prevent reading and returning data that is not needed. If you need all columns, it's valid to pass in null. You provide a string array of columns to return.

- `selection`: A filter describing what rows to return formatted as an SQL `WHERE` clause (excluding the `WHERE` itself). Passing a null returns all rows in the table. Depending on the situation, you provide either the `rowId` of the task you would like to fetch or you provide a null to return all tasks.

- `selectionArgs`: You may include question marks (?) in the selection, which are replaced by the values from `selectionArgs` in the order they appear in the selection. These values are bound as string types. You do not need `selectionArgs`; therefore, you pass in null.

- `groupBy`: A filter describing how to filter rows formatted as an SQL `GROUP BY` clause (excluding the `GROUP BY`). Passing null causes the rows not to be grouped. You pass a null value because you don't care how the results are grouped.

- `having`: This filter describes row groups to include in the cursor, if row grouping is being used. Passing null causes all row groups to be included, and is required when row grouping is not being used. You pass in a null value.

- `orderBy`: How to order the rows, formatted as an SQL `ORDER BY` clause (excluding the `ORDER BY` itself). Passing null uses the default sort order, which may be unordered. You pass in a null value because you aren't concerned with the order in which the results are returned.

- `limit`: Limits the number of rows returned by the query by utilizing a `LIMIT` clause. Passing null states that you do not have a `LIMIT` clause. You don't want to limit the number of rows returned; therefore, you pass in null to return all the rows that match your query.

Understanding the update operation

Updating a record in a database simply takes the incoming parameters and replaces them in the destination cell inside the row specified (or in the rows if many rows are updated). As with the following delete operation, the update can affect many rows. It is important to understand the update method's parameters and how they can affect the records in the database. The `update()` method accepts the following parameters:

- ✔ `table`: The table to update. The value that you use is provided by the `DATABASE_TABLE` constant.

- ✔ `values`: The `ContentValues` object, which contains the fields to update. Use the `args` variable, which you constructed on line 70 of Listing 14-4.

- ✔ `whereClause`: The `WHERE` clause, which restricts which rows should get updated. Here you inform the database to update the row whose ID is equal to `rowId` by providing the following string value: `KEY_ROWID + "=" + rowId`.

- ✔ `whereArgs`: Additional `WHERE` clause arguments. Not used in this call; therefore, null is passed in.

Understanding the delete operation

When using the `delete()` method, various parameters are used to define the deletion criteria in the database. A delete statement can affect none or all of the records in the database. It is important to understand the parameters of the delete call to ensure that you do not mistakenly delete data. The parameters for the `delete()` method are as follows:

- ✔ `table`: The table to delete the rows from. The value of this parameter is provided by the `DATABASE_TABLE` constant.

- ✔ `whereClause`: This is the optional `WHERE` clause to apply when deleting rows. If you pass null, all rows will be deleted. This value is provided by manually creating the `WHERE` clause with the following string: `KEY_ROWID + "=" + rowId`.

- ✔ `whereArgs`: The optional `WHERE` clause arguments. Not needed in this call because everything is provided through the `WHERE` clause itself. You pass in a null value because you do not need to use this parameter.

Returning all the tasks with a cursor

You can create a task, but what good is it if you can't see the task in the task list? No good at all, really. Therefore, you must list the tasks that currently exist in the database in the `ListView` in the `ReminderListActivity`.

Listing 14-5 outlines the entire `ReminderListActivity` with the new code that can read the list of tasks from the database into the `ListView`.

Listing 14-5:   The Entire ReminderListActivity with Connections to SQLite

```
public class ReminderListActivity extends ListActivity {
    private static final int ACTIVITY_CREATE=0;
    private static final int ACTIVITY_EDIT=1;

    private RemindersDbAdapter mDbHelper;                              →5

    /** Called when the activity is first created. */
    @Override
    public void onCreate(Bundle savedInstanceState) {
        super.onCreate(savedInstanceState);
        setContentView(R.layout.reminder_list);
        mDbHelper = new RemindersDbAdapter(this);
        mDbHelper.open();
        fillData();                                                   →14
        registerForContextMenu(getListView());

    }

    private void fillData() {
        Cursor remindersCursor = mDbHelper.fetchAllReminders();       →20
        startManagingCursor(remindersCursor);                         →21

        // Create an array to specify the fields we want (only the TITLE)
        String[] from = new String[]{RemindersDbAdapter.KEY_TITLE};   →24

        // and an array of the fields we want to bind in the view
        int[] to = new int[]{R.id.text1};                             →27

        // Now create a simple cursor adapter and set it to display
        SimpleCursorAdapter reminders =
                new SimpleCursorAdapter(this, R.layout.reminder_row,
                    remindersCursor, from, to);                       →32
        setListAdapter(reminders);                                    →33
    }

    // Menu Code removed for brevity

    @Override
    protected void onListItemClick(ListView l, View v, int position, long id) {
        super.onListItemClick(l, v, position, id);
        Intent i = new Intent(this, ReminderEditActivity.class);
        i.putExtra(RemindersDbAdapter.KEY_ROWID, id);                 →42
        startActivityForResult(i, ACTIVITY_EDIT);
    }
```

(continued)

Listing 14-5 *(continued)*

```
    @Override
    protected void onActivityResult(int requestCode, int resultCode, Intent
            intent) {
        super.onActivityResult(requestCode, resultCode, intent);
        fillData();                                                        →49
    }


    @Override
    public boolean onContextItemSelected(MenuItem item) {                  →53
        switch(item.getItemId()) {
            case R.id.menu_delete:
                AdapterContextMenuInfo info =
                        (AdapterContextMenuInfo)  item.getMenuInfo();       →57
                mDbHelper.deleteReminder(info.id);                         →58
                fillData();                                                →59
                return true;
        }
        return super.onContextItemSelected(item);
    }

}
```

The code for reading the list of tasks is explained as follows:

→**5**   This line of code defines a class-level RemindersDbAdapter instance variable. The variable is instantiated in the onCreate() method.

→**14**   The fillData() method is called, which loads the data from the SQLite database into the ListView.

→**20**   When you're inside the fillData() method, you fetch all the reminders from the database, as shown on line 52 of Listing 14-4.

→**21**   This line uses the manage startManagingCursor() method, which is present on the Activity class. This method allows the activity to take care of managing the given Cursor's life cycle based on the activity's life cycle. For example, when the activity is stopped, the activity automatically calls deactivate() on the Cursor, and when the activity is later restarted, it calls requery() for you. When the activity is destroyed, all managed Cursors are closed automatically.

→**24**   On this line, you define the selection criteria for the query. You request that the task title be returned.

→**27**   On this line, you define the array of views that you want to bind to as the view for the row. Therefore, when you're displaying a task title, that title will correspond to a particular task ID. This is why the variable in line 24 is named from and the variable on this line is named to. The values from line 24 map to the values on line 27.

→**32** On this line, you create a `SimpleCursorAdapter` that maps columns from a `Cursor` to `TextViews` as defined in a layout XML file. By using this method, you can specify which columns you want to display and the XML file that defines the appearance of these views. The use of a `SimpleCursorAdapter` and the associated parameters is described in the following section.

→**33** The `SimpleCursorAdapter` is passed as the adapter parameter to the `setListAdapter()` method to inform the list view where to find its data.

→**42** This line of code places the `ID` of the task to be edited into the intent. The `ReminderEditActivity` inspects this intent, and if it finds the `ID`, it attempts to allow the user to edit the task.

→**49** The `fillData()` method is called when the activity returns from another activity. This is called here because the user might have updated or added a new task. Calling this method ensures that the new task is present in the list view.

→**53** This line defines the method that handles the user context menu events that occur when a user selects a menu item from the context menu after a long press on the task in the list view.

→**57** This line of code uses the `getMenuInfo()` method of the item that was clicked to obtain an instance of `AdapterContextMenuInfo`. This class exposes various bits of information about the menu item and item that was long-pressed in the list view.

→**58** This line of code calls into the `RemindersDbAdapter` to delete the task whose ID is retrieved from the `AdapterContextMenuInfo` object's `id` field. This `id` field contains the ID of the row in the list view. This ID is the `rowId` of the task in the database.

→**59** After the task has been deleted from the system, you call `fill-Data()` to repopulate the task list. This refreshes the list view, removing the deleted item.

Understanding the SimpleCursorAdapter

In line 32 of Listing 14-5, you created a `SimpleCursorAdapter`. The `SimpleCursorAdapter` does a lot of the hard work for you when you want to bind data from a `Cursor` object to a list view. To set up a `SimpleCursorAdapter`, you need to provide the following parameters:

✔ `this: Context`: The context that is associated with the adapter.

✔ `R.layout.reminder_row - layout`: The layout resource identifier that defines the file to use for this list item.

✔ reminderCursor - c: The database Cursor.

✔ from - from: An array of column names that are used to bind data from the cursor to the view. This is defined on line 24.

✔ to - to: An array of view IDs that should display the column information from the from parameter. The To field is defined on line 27.

✔ The to and from parameters create a mapping informing the SimpleCursorAdapter how to map data in the cursor to views in the row layout.

Now, when you start the application, you see a list of items that you have created. These items are being read from the SQLite database. If you do not see a list of items, create one by pressing the Add Reminder button in the action bar that allows you to add a new task.

Deleting a task

To the end user, deleting a task is as simple as long-pressing an item in the ReminderListActivity and selecting the delete action, but to actually delete the task from the database, you need to use the delete() method on the SQLite database object. This method is called in Listing 14-4 on line 49.

The RemindersDbAdapter deleteReminder() method is called from within the onContextSelectedItem() method call on line 58 of Listing 14-5. The one item that is needed prior to deleting the task from the database is the rowId of the task in the database. To obtain the rowId, you must use the AdapterContextMenuInfo object, which provides extra menu information. This information is provided to the context menu selection when a menu is brought up for the ListView. Because you load the list with a database cursor, the ListView contains the rowId that you're looking for — yes, it's that simple! On line 57 of Listing 14-5, you obtain the AdapterContextMenuInfo object, and on line 58, you call the delete() method with the rowId as a parameter. Afterward, you call the fillData() method to reload the tasks to the screen. You can now create, list (read), and delete the task. The only thing left is updating the task.

Updating a task

When it comes down to it, updating a task is a fairly trivial process. However, it can get a bit tricky because you must use the same activity to update a task as you used to create the task. Therefore, logic has to be put into place to determine whether you are editing an existing task or creating a new one.

This logic is based on the intent that was used to start the activity. In the `ReminderListActivity`, when an item is tapped, the following activity is started:

```
Intent i = new Intent(this, ReminderEditActivity.class);
i.putExtra(RemindersDbAdapter.KEY_ROWID, id);
startActivityForResult(i, ACTIVITY_EDIT);
```

This code informs Android to start the `ReminderEditActivity` with the i parameter (the intent), which contains extra information — the row id of the task that you would like to edit. On the `ReminderEditActivity` side, you inspect the receiving intent to determine whether it contains the extra id information. If it does, you then consider this an edit action and load the task information into the form to allow the user to edit the information. If the extra information is not there (which would happen if the user elected to add a new task from the menu), you present the user with an empty form to fill out to create a new task.

See Listing 14-6 for an implementation of the previously described logic. The bolded sections outline the new code.

Listing 14-6: The ReminderEditActivity That Supports Inserting and Updating a Task

```java
public class ReminderEditActivity extends Activity {

    // Other Class level variables go here. Removed for brevity
    private Long mRowId;

    @Override
    protected void onCreate(Bundle savedInstanceState) {
        super.onCreate(savedInstanceState);

        mDbHelper = new RemindersDbAdapter(this);

        setContentView(R.layout.reminder_edit);

        mCalendar = Calendar.getInstance();
        mTitleText = (EditText) findViewById(R.id.title);
        mBodyText = (EditText) findViewById(R.id.body);
        mDateButton = (Button) findViewById(R.id.reminder_date);
        mTimeButton = (Button) findViewById(R.id.reminder_time);

        mConfirmButton = (Button) findViewById(R.id.confirm);

        mRowId = savedInstanceState != null                          →22
            ? savedInstanceState.getLong(RemindersDbAdapter.KEY_ROWID)
            : null;
```

(continued)

Listing 14-6 *(continued)*

```
         registerButtonListenersAndSetDefaultText();
    }

    private void setRowIdFromIntent() {                                    →28
        if (mRowId == null) {
            Bundle extras = getIntent().getExtras();
            mRowId = extras != null
                ? extras.getLong(RemindersDbAdapter.KEY_ROWID)
                : null;
        }
    }

    @Override
    protected void onPause() {
        super.onPause();
        mDbHelper.close();                                                 →40
    }

    @Override
    protected void onResume() {                                            →44
        super.onResume();
        mDbHelper.open();                                                  →46
        setRowIdFromIntent();                                             →47
        populateFields();                                                →48
    }

    // Date picker, button click events, and buttonText updating, createDialog
    // left out for brevity
    // they normally go here ...

    private void populateFields() {                                       →55
        if (mRowId != null) {                                             →57
            Cursor reminder = mDbHelper.fetchReminder(mRowId);
            startManagingCursor(reminder);                                →58
            mTitleText.setText(reminder.getString(
        reminder.getColumnIndexOrThrow(RemindersDbAdapter.KEY_TITLE)));   →60
                mBodyText.setText(reminder.getString(
        reminder.getColumnIndexOrThrow(RemindersDbAdapter.KEY_BODY)));    →62
            SimpleDateFormat dateTimeFormat =
                    new SimpleDateFormat(DATE_TIME_FORMAT);               →64
            Date date = null;                                             →65
            try {
                String dateString =    reminder.getString(
                    reminder.getColumnIndexOrThrow(
                            RemindersDbAdapter.KEY_DATE_TIME));           →69
                date = dateTimeFormat.parse(dateString);                 →70
                mCalendar.setTime(date);                                 →71
            } catch (ParseException e) {                                 →72
                Log.e("ReminderEditActivity", e.getMessage(), e);       →73
            }
        }
```

```
                updateDateButtonText();
        updateTimeButtonText();
    }


    @Override
    protected void onSaveInstanceState(Bundle outState) {
        super.onSaveInstanceState(outState);
        outState.putLong(RemindersDbAdapter.KEY_ROWID, mRowId);          →84
    }

    private void saveState() {
        String title = mTitleText.getText().toString();
        String body = mBodyText.getText().toString();

        SimpleDateFormat dateTimeFormat = new SimpleDateFormat(DATE_TIME_
            FORMAT);
            String reminderDateTime =
                    dateTimeFormat.format(mCalendar.getTime());

        if (mRowId == null) {                                           →96
            long id = mDbHelper.createReminder(title, body, reminderDateTime);
                →97
            if (id > 0) {                                               →98
                mRowId = id;                                            →99
            }
        } else {
            mDbHelper
                .updateReminder(mRowId, title, body, reminderDateTime);  →103
        }
    }
}
```

Each line of code is explained as follows:

→22   The instance state is checked to see whether it contains any values for the mRowId. The instance state is set on line 84.

→28   This method sets the mRowId from the intent that started the activity. If the Intent object does not contain any extra information, the mRowId object is left null. Note that you use a Long (with a capital *L*). This is a reference-type long — meaning that this object can be null or it can contain a long value.

→40   Before the activity is shut down or when it's paused, the database is closed.

→44   The onResume() method is called as part of the normal activity life cycle. This life cycle is explained in Chapter 7.

→46   The database is opened so that you can use it in this activity.

→47   This method call sets the mRowId object from the intent that started the activity.

→**48**   The `populateFields()` method is called to populate the form.

→**55**   This method populates the form if the `mRowId` object is not null.

→**57**   This line of code retrieves a `Cursor` from the SQLite database based on the `mRowId`. The `fetchReminder()` call is made on line 57 of Listing 14-4.

→**58**   This line starts the activity management of the `Cursor`.

→**60**   This line sets the text of the title using the `Cursor`. To retrieve values from the cursor, you need to know the index of the column in the cursor. The `getColumnIndexOrThrow()` method on the `Cursor` object provides the column index when given the column name. After the column index is retrieved, you can obtain the column value by calling `getString()` with the column index as a parameter. After the value is retrieved, you set the text of the `mTitleText EditText` view.

→**62**   This line retrieves and sets the value for the `mBodyTest EditText` view using the same method as described in line 60, but this time it uses a different column name and index.

→**64**   Because SQLite does not store actual date types, they are stored as strings. Therefore, you need to create a `SimpleDateFormat` to parse the date. This is the `SimpleDateFormat` that parses the date from the database.

→**65**   This line instantiates a new `Date` object from the `java.util.Date` package.

→**69**   This line retrieves the date as a string from the `Cursor`.

→**70**   This line parses the date into a `Calendar` object.

→**71**   This line sets the calendar object's date and time from the date and time that was parsed from the database.

→**72**   This line catches any parse exceptions that may occur due to incorrect string formats that are passed into the `SimpleDateFormat` parsing. The `ParseException` that is caught here is from the `java.text.ParseException` package.

→**73**   This line prints the error message to the system log.

→**84**   This line saves the `mRowId` instance state. The `onSaveInstanceState()` method is called so that you may retrieve and store activity-level instance states in a `Bundle`. This method is called before the activity is killed so that when the activity comes back in the future, it can be restored to a known state (as done in the `onResume()` method). On line 22, you check to see whether a `rowId` is present in the `savedInstanceState` object prior to checking the intent for incoming data. You do this because there may be a point in time when Android kills the activity for some reason while you're using the app. Such instances include, but are not limited to, using the Maps feature for navigation, playing music, and so on. At a later time, when you finally return to the app, the savedInstanceState can be inspected to see whether the activity can resume what it was doing before. Storing the mRowId in this object allows me to resume working with the activity in a pre-determined state.

→96   In the saveState() method, you determine whether to save a new task or update an existing one. If the mRowId is null, that means that no row Id could be found in the savedInstanceState or in the incoming intent; therefore, the task is considered new.

→97   A new task is created in the database.

→98   Checks to make sure that the ID returned from the insert is greater than zero. All new inserts return their ID, which should be greater than zero.

→99   Setting the local mRowId to the newly created ID.

→103   This line updates the task. You pass in the row Id to update the title, the body, and the reminder date and time to update the task with.

When you fire up the application in the emulator, you can now create, read, update, and delete tasks! The only things left to build are the reminders' status bar notifications!

Chapter 15

Reminding the User with AlarmManager

Many tasks need to happen on a daily basis, right? Wake up, take a shower, eat breakfast, and so on — I'm sure they all sound familiar. It's everyone's Monday-through-Friday prework morning routine. You maybe have an internal clock that gets you up every day on time, but I have to set alarms to ensure I get to work on time! At work, a calendar reminds me of upcoming events — such as meetings and important server upgrades. Reminders and alarms are part of everyone's everyday routine, and we all rely on them in one way or another.

Building your own scheduled task system is a pain. Thankfully, you don't have to — Android has the `AlarmManager` class to help you with that.

Your Task Reminder application must be able to remind users of their tasks, according to the schedule they set. This is where the `AlarmManager` class comes into play.

The `AlarmManager` class allows you to specify a time for your application to run. When an alarm goes off, an intent is broadcast by the system. Your application then responds to that broadcast intent and performs an action — such as opening your application and notifying the user by means of a status bar notification (which you do in Chapter 16) or performing some other type of action.

In this chapter, you work with `AlarmManager` to add scheduling functionality to your Task Reminder application.

Waking Up a Process with AlarmManager

To wake up a process with the `AlarmManager`, you have to set the alarm first. In the Task Reminder application, the best place to do that is right after you save your task in the `saveState()` method in the `ReminderEditActivity`. Before you add that code, however, you need to add four class files to your project in the `src/` folder:

✔ `ReminderManager.java`: This class is responsible for setting up reminders using the `AlarmManager`. The code for this class is shown in Listing 15-1 (see the next section in this chapter).

✔ `OnAlarmReceiver.java`: This class is responsible for handling the broadcast when the alarm goes off. The code for this class is shown in Listing 15-2 (see the section "Creating the OnAlarmReceiver class," later in this chapter). In addition to adding the code from that section, you also need to add the following line of code to the application element in your `AndroidManifest.xml` file for your application to recognize this receiver:

```
<receiver android:name=".OnAlarmReceiver" />
```

The leading period syntax informs Android that the receiver is in the current package — the one that is defined in the application element of the `ApplicationManifest.xml` file.

✔ `WakeReminderIntentService.java`: This abstract class is responsible for acquiring and releasing the wake lock. The code for this class is shown in Listing 15-3 (see the section "Creating the WakeReminderIntentService class," later in this chapter).

✔ `ReminderService.java`: This class is an implementation of the `WakeReminderIntentService` that handles the building of the notification, as shown in Chapter 14. The code for this class is shown in Listing 15-4 (see the section "Creating the ReminderService class," later in this chapter).

In addition to adding this code, you also need to add the following line of code to the application element in the `AndroidManifest.xml` file for your application to recognize this service:

```
<service android:name=".ReminderService" />
```

Creating the ReminderManager class

As stated previously, the `ReminderManager` class is responsible for setting up alarms using the `AlarmManager` class in Android. You place all actions pertaining to setting alarms from the `AlarmManager` into this class.

Add the following code to the end of the saveState() method in the ReminderEditActivity class to add an alarm for that task:

```
new ReminderManager(this).setReminder(mRowId, mCalendar);
```

This line of code instructs the ReminderManager to set a new reminder for the task with a row ID of mRowId at the particular date and time as defined by the mCalendar variable.

Listing 15-1 shows the code for the ReminderManager class.

Listing 15-1:   ReminderManager Class

```
public class ReminderManager {

private Context mContext;
private AlarmManager mAlarmManager;
public ReminderManager(Context context) {                              →5
    mContext = context;
    mAlarmManager =
        (AlarmManager)context.getSystemService(Context.ALARM_SERVICE);  →8
}


public void setReminder(Long taskId, Calendar when) {                  →11
    Intent i = new Intent(mContext, OnAlarmReceiver.class);           →12
    i.putExtra(RemindersDbAdapter.KEY_ROWID, (long)taskId);           →13

    PendingIntent pi =
    PendingIntent.getBroadcast(mContext, 0, i,
                    PendingIntent.FLAG_ONE_SHOT);                      →16

    mAlarmManager.set(AlarmManager.RTC_WAKEUP, when.getTimeInMillis(), pi); →18
        }
}
```

Each numbered line of code is explained as follows:

→5   The ReminderManager class is instantiated with a context object.

→8   An AlarmManager is obtained through the getSystemService() call.

→11   The setReminder() method is called with the database ID of the task and the Calendar object of when the alarm should fire.

→12   A new Intent object is created. This intent object is responsible for specifying what should happen when the alarm goes off. In this instance, I am specifying that the OnAlarmReceiver receiver should be called.

→**13**   The `Intent` object is provided with extra information — the ID of the task in the database.

→**16**   The `AlarmManager` operates in a separate process, and for the `AlarmManager` to notify an application that an action needs to be performed, a `PendingIntent` must be created. The `PendingIntent` contains an `Intent` object that was created on line 13. On this line, a `PendingIntent` is created with a flag of `FLAG_ONE_SHOT` to indicate that this `PendingIntent` can only be used once.

→**18**   The `AlarmManager`'s `set()` method is called to schedule the alarm. The `set()` method is provided with the following parameters:

- type: `AlarmManager.RTC_WAKEUP`: Wall-clock time in UTC. This parameter wakes up the device when the specified `triggerAt Time` argument time elapses.

- `triggerAtTime: when.getTimeInMillis()`: The time the alarm should go off. The `Calendar` object provides the `get TimeInMillis()` method, which converts the time into long value, which represents time in units of milliseconds.

- operation: `pi`: The pending intent to act upon when the alarm goes off. The alarm will now go off at the time requested.

If an alarm is already scheduled with a pending intent that contains the same signature, the previous alarm will first be canceled and a new one will be set up.

Creating the OnAlarmReceiver class

The `OnAlarmReceiver` class (see Listing 15-2) is responsible for handling the intent that is fired when an alarm is raised. This class acts as a hook into the alarm system because it is essentially a simple implementation of a `BroadcastReceiver` — which can react to broadcast events in the Android system.

Listing 15-2:   OnAlarmReceiver Class

```
public class OnAlarmReceiver extends BroadcastReceiver {
    @Override
    public void onReceive(Context context, Intent intent) {
        long rowid =
            intent.getExtras().getLong(RemindersDbAdapter.KEY_ROWID);    →5

        WakeReminderIntentService.acquireStaticLock(context);    →7

        Intent i = new Intent(context, ReminderService.class);    →9
```

```
        i.putExtra(RemindersDbAdapter.KEY_ROWID, rowid);      →10
        context.startService(i);                              →11

    }
}
```

Each numbered line is explained as follows:

→**5**   I am retrieving the database ID of the task from the intent after the receiver has started handling the intent.

→**7**   Here, I inform the `WakeReminderIntentService` to acquire a static lock on the CPU to keep the device alive while work is being performed.

→**9**   This line defines a new `Intent` object that will start the `ReminderService`.

→**10**   On this line, I am placing the ID of the task into the intent that will be used to start the service that will do the work. This gives the `ReminderService` class the ID of the task that it needs to work with.

→**11**   This line starts the `ReminderService`.

This is the first entry point for the alarm you set. In this `BroadcastReceiver`, you would not want to let the device go back to sleep during your processing because your task would never complete and could possibly leave your application in a broken state through data corruption with the database.

When an alarm goes off, the pending intent that was scheduled with the alarm is broadcast through the system, and any broadcast receiver that is capable of handling it will handle it.

Because this is your second foray into the `BroadcastReceiver` object, you're probably still a bit fuzzy about how they work. A `BroadcastReceiver` is a component that does nothing but receive and react to system broadcast messages. A `BroadcastReceiver` does not display a user interface; however, it starts an activity in response to the broadcast. The `OnAlarmReceiver` is an instance of a `BroadcastReceiver`.

When the `AlarmManager` broadcasts the pending intent, the `OnAlarmReceiver` class responds to the intent — because it is addressed to that class as shown on line 12 of Listing 15-1. This class then accepts the intent, locks the CPU, and performs the necessary work.

Creating the WakeReminder IntentService class

The `WakeReminderIntentService` class is the base class for the `ReminderService` class, as shown in Listing 15-3. This class handles the management of acquiring and releasing a CPU wake lock. A CPU wake lock keeps the device on (but not necessarily the screen) while some work takes place. After the work is complete, this class releases the wake lock so that the device may return to sleep.

The `AlarmManager` holds a CPU wake lock as long as the alarm receiver's `onReceive()` method is executing. A wake lock guarantees that the tablet will not sleep until you have finished working with the broadcast. This is the reason why you needed the `WAKE_LOCK` permission that was set up in the previous chapter.

Listing 15-3: WakeReminderIntentService Class

```
public abstract class WakeReminderIntentService extends IntentService {
        abstract void doReminderWork(Intent intent);                        →2

    public static final String
        LOCK_NAME_STATIC="com.dummies.android.taskreminder.Static";          →5
    private static PowerManager.WakeLock lockStatic=null;                    →6

    public static void acquireStaticLock(Context context) {
        getLock(context).acquire();                                          →9
    }

    synchronized private static PowerManager.WakeLock
                            getLock(Context context) {                       →13
        if (lockStatic==null) {
            PowerManager
                mgr=(PowerManager)context
                    .getSystemService(Context.POWER_SERVICE);                →17

          lockStatic=mgr.newWakeLock(PowerManager.PARTIAL_WAKE_LOCK,
                            LOCK_NAME_STATIC);                                →20
            lockStatic.setReferenceCounted(true);                           →21
    }

        return(lockStatic);                                                  →23
    }

    public WakeReminderIntentService(String name) {                         →26
        super(name);
    }
```

```
        @Override
        final protected void onHandleIntent(Intent intent) {          →31
            try {
                doReminderWork(intent);                                →33
            } finally {
                getLock(this).release();                               →35
            }

        }
}
```

Each numbered line is explained as follows:

→2   This abstract method is implemented in any children of this class — such as in the child `ReminderService` as shown on line 7 of Listing 15-4.

→5   This line is the tag name of the lock that you will use to acquire the CPU lock. This tag name assists in debugging.

→6   This line is the private static wake lock variable, which is referenced and set later in this class.

→9   This line calls the `getLock()` method, as described on line 13. After that call is returned, the `acquire()` method is called to ensure that the device is on in the state that you requested, a partial wake lock. This wake lock prevents the device from sleeping, but it doesn't turn on the screen.

→13   This line defines the `getLock()` method that returns the `PowerManager.WakeLock`, which lets you inform Android that you would like the device to stay on to do some work.

→17   This line retrieves the `PowerManager` from the `getSystemService()` call. This is used to create the lock.

→20   This line creates a new `WakeLock` using the `newWakeLock()` method call. This method accepts the following parameters:

- `flags`: `PowerManager.PARTIAL_WAKE_LOCK`: You can provide numerous tags to this call; however, I am only providing this single tag. The `PARTIAL_WAKE_LOCK` tag informs Android that you need the CPU to be on, but the screen does not have to be on.

- `tag`: `LOCK_NAME_STATIC`: The name of your class name or another string. This is used for debugging purposes. This is a custom string that is defined on line 5.

→21   This line informs the `PowerManager` that this reference has been counted.

→**23**   This line returns the `WakeLock` to the caller.

→**26**   This line is the constructor with the name of the child instance that has created it. This name is used for debugging only.

→**31**   This line is the `onHandleIntent()` call of the `IntentService`. As soon as the service is started, this method is called to handle the intent that was passed to it.

→**33**   The service attempts to perform the necessary work by calling `doReminderWork()`.

→**35**   Regardless of whether the call to `doReminderWork()` is successful, you want to make sure that you release the `WakeLock`. If you do not release the `WakeLock`, the device could be left in an On state until the phone is rebooted, which ultimately would drain the battery. For this reason, the `release()` method is called in the final portion of the `try-catch` block. The final portion of the `try-catch` block is always called, regardless of whether the try succeeds or fails.

Although no implementation for the `doReminderWork()` exists in the `ReminderService` just yet, the Task Reminder application responds to alarms. Feel free to set up multiple tasks and to set break points in the debugger to watch the execution path break in the `ReminderService do ReminderWork()` method.

The `AlarmManager` does not persist alarms. This means that if the device gets rebooted, the alarms must be set up again. Each time the phone is rebooted, the alarms need to be set up again.

The previous code demonstrates what is necessary to perform work on a device that might be asleep or locked. This code acquires the wake lock, and while the device is locked into a wakeful state, you call into `doReminder-Work()`, which is implemented in the `ReminderService`.

Creating the ReminderService class

The `ReminderService` class (see Listing 15-4) is responsible for doing the work when an alarm is fired. The implementation in this chapter simply creates a shell for work to take place. You implement the status bar notification in Chapter 16.

Listing 15-4: ReminderService Class

```
public class ReminderService extends WakeReminderIntentService {    →1
    public ReminderService() {
        super("ReminderService");
    }

    @Override
    void doReminderWork(Intent intent) {                              →7
        Long rowId = intent.getExtras()
            .getLong(RemindersDbAdapter.KEY_ROWID);                   →9

        // Status bar notification Code Goes here.
    }
}
```

Each numbered line of code is explained as follows:

→1   This line defines the `ReminderService` class by inheriting from the `WakeReminderIntentService`.

→7   The abstract method `doReminderWork()` in the `WakeReminderIntentService` is implemented here.

→9   On this line, you retrieve the task ID inside the `Intent` object that passed in this class.

As noted before, this class contains no implementation — other than retrieving the ID of the task from the intent.

Rebooting Devices

I admit, after a long day and a good night's rest, I forget things from time to time. I'm only human, right? I usually have to be reminded of certain things when I wake up; that's just the way it is. The Android `AlarmManager` is no different. The `AlarmManager` does not persist alarms; therefore, when the device reboots, you must set up the alarms all over again. Although it's not a huge pain in the butt, it's something worth knowing.

If you do not set up your alarms again, they simply will not fire, because to Android they do not exist.

Creating a boot receiver

In the last chapter, you set up the RECEIVE_BOOT_COMPLETED permission. This permission allows your application to receive a broadcast notification from Android when the device is done booting and is eligible to be interactive with the user. Because the Android system can broadcast a message when this event is complete, you need to add another BroadcastReceiver to your project. This BroadcastReceiver is responsible for handling the boot notification from Android. When the broadcast is received, the receiver needs to connect to SQLite through the RemindersDbAdapter, loop through each task, and schedule an alarm for it. This procedure ensures that your alarms don't get lost in the reboot.

Add a new BroadcastReceiver to your application. For the Task Reminder application, I'm giving it a name of OnBootReceiver. You also need to add the following lines of code to the application element in the AndroidManifest.xml file:

```
<receiver android:name=".OnBootReceiver">
    <intent-filter>
        <action android:name="android.intent.action.BOOT_COMPLETED" />
    </intent-filter>
</receiver>
```

This informs Android that the OnBootReceiver should receive boot notifications for the BOOT_COMPLETED action. In layman's terms — let OnBootReceiver know when the device is done booting up.

The full implementation of OnBootReceiver is shown in Listing 15-5.

Listing 15-5: OnBootReceiver

```
public class OnBootReceiver extends BroadcastReceiver {                      →1

    @Override
    public void onReceive(Context context, Intent intent) {                 →4
        ReminderManager reminderMgr = new ReminderManager(context);         →5
        RemindersDbAdapter dbHelper = new RemindersDbAdapter(context);
        dbHelper.open();
        Cursor cursor = dbHelper.fetchAllReminders();                       →8
        if(cursor != null) {
            cursor.moveToFirst();                                           →10

            int rowIdColumnIndex = cursor.getColumnIndex(RemindersDbAdapter.
              KEY_ROWID);
                int dateTimeColumnIndex =
                cursor.getColumnIndex(RemindersDbAdapter.KEY_DATE_TIME);
```

```
        while(cursor.isAfterLast() == false) {                   →16
          Long rowId = cursor.getLong(rowIdColumnIndex);
          String dateTime = cursor.getString(dateTimeColumnIndex);

          Calendar cal = Calendar.getInstance();
          SimpleDateFormat format = new
              SimpleDateFormat(ReminderEditActivity.DATE_TIME_FORMAT);

      try {
            java.util.Date date = format.parse(dateTime);        →25
            cal.setTime(date);                                   →26

            reminderMgr.setReminder(rowId, cal);                 →28
      } catch (ParseException e) {
            Log.e("OnBootReceiver", e.getMessage(), e);          →30
      }

      cursor.moveToNext();                                       →33
    }
    cursor.close() ;                                             →35
  }

      dbHelper.close();                                          →38
  }
}
```

Each numbered line is explained in detail as follows:

→**1**   This line is the definition of the `OnBootReceiver`.

→**4**   This line is the `onReceive()` method that is called when the receiver receives an intent to perform an action.

→**5**   This line sets up a new `ReminderManager` object that allows me to schedule alarms.

→**8**   This line obtains a cursor with all the reminders from the `RemindersDbAdapter`. This call is also used to load the `ListView` in the `ReminderListActivity`.

→**10** This line moves to the first record in the `Cursor`. Because a cursor can contain many records, you can advance the cursor to the next record upon request. That is what you are doing here.

→**16** This line sets up a `while` loop. This `while` loop checks to see whether the cursor is moved past the last record. If it equals false, this means that you are still working with a valid record. You move the cursor to the next record on line 33. If this value were true, it would mean that no more records were available to utilize in the cursor.

→**25** The date is parsed from the string retrieved from the database.

→**26** After the date is retrieved from the cursor, the `Calendar` variable needs to be updated with the correct time. This line formats the parsed date value into the local `Calendar` object.

→**28** This line schedules a new reminder with the row ID from the database at the time defined by the recently built `Calendar` variable.

→**30** This line prints any exceptions to the system log.

→**33** This line moves to the next record in the cursor. If no more records exist in the cursor, the call to `isAfterLast()` returns true, which means that the `while` loop will exit. After this line executes, the loop processes again by returning execution to line 16 and continuing the process until no more database records are left.

→**35** This line closes the cursor because it is no longer needed. When you previously worked with the `Cursor` object, you may have noticed that you never had to close the cursor. This is because the `Activity` object was managing the cursor for me. Because you're in a broadcast receiver, you do not have access to the `Activity` class because it is not in scope and is not valid in this instance.

→**38** This line closes the `RemindersDbAdapter`, which in turn closes the database because it is no longer needed.

If you were to start the application, create a few reminders, and then reboot the device, you would now see that the reminders persisted. If you decide to debug the application, be sure to set the debuggable attribute to true in the application manifest.

Checking the boot receiver

If you're not sure whether the `OnBootReceiver` is working, place log statements into the `while` loop like this:

```
Log.d("OnBootReceiver", "Adding alarm from boot.");
Log.d("OnBootReceiver", "Row Id Column Index - " + rowIdColumnIndex);
```

This prints messages to the system log that are viewable through DDMS. You can then shut down the emulator (or device) and then start it again. Watch the messages stream through in DDMS and look for the `OnBootReceiver` messages. If you have two tasks in your database, you should see two sets of messages informing you of the system adding an alarm from boot. Then the next message should be the row ID column index.

Chapter 16

Updating the Android Status Bar

Throughout this book, I cover various ways to grab the user's attention, including dialog boxes, toasts, and new activities. Although these techniques work well in their respective situations, sometimes you need to inform the user of something without stealing his attention from the current activity. That is, you need a way to inform the user that something needs his attention — but only when he has time to tend to the matter. This is exactly what the status bar is for.

In this chapter, I show you how to use the status bar to communicate messages to the user, and I show you how to add this functionality to your Task Reminder app.

Deconstructing the Status Bar

Let me resort to the age-old saying of a picture being worth a thousand words: The best way to describe the status bar is to show it to you. See Figure 16-1.

Viewing status bar icons

In Figure 16-1, the first icon at the bottom right is the Lookout security application's way of informing me that my tablet is currently secure. The second icon tells me that the device is connected to another device (a computer) via USB, and the third icon informs me that USB debugging is enabled. By pressing the status bar near the clock, I can receive more information, as shown in Figure 16-2.

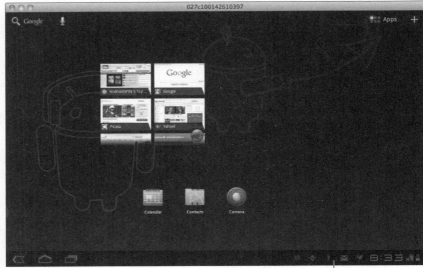

Figure 16-1:
The status
bar with
multiple
icons
present.

Status bar

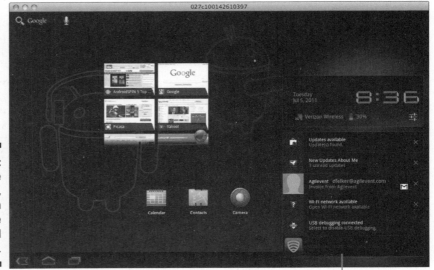

Figure 16-2:
Opening the
status bar,
also known
as the
expanded
view.

Status bar open

In Figure 16-2, you can see that each notification has an expanded view that gives you more information. The user can select the expanded view that interests her — which starts the requested action.

As a developer, you have access to modify the contents of the status bar.

Using status bar tools to notify the user

The status bar's icons aren't the only way for you to notify the user. You can augment your notification with additional flags (which I cover later in this chapter) during the notification process. Some of these options are as follows:

- **Sound:** Sound some type of alarm when the notification occurs, such as a notification sound or prerecorded tone that you install with your application. This option is useful if the user has the notification sound level cranked up.

- **Lights:** Many devices contain an LED that you have programmatic access to. You can tell the light to flash at a given interval with a specific color that you program. If the LED only supports one color (such as white), it will flash white, ignoring your color requirement. If the user has the device set to silent, using lights provides an excellent cue that something needs attention.

Adding these various options to your notification arsenal can help immensely because they let the user know that something has happened on the device.

The status bar is a very powerful tool because it can be used to provide valuable feedback to the user throughout the lifetime of an application. Although icons, vibration, lights, and sound might sound like a golden jackpot, that's not the end of the rainbow, Mr. Leprechaun. Notifications also allow you to provide scrolling information to the user. This is the information that shows when the notification first arrives. After that, the user needs to tap the status bar icon to see the expanded view (as shown in Figure 16-2).

The status bar framework can be used to inform users of a number of activities, including device state, new mail notifications, and even progress downloads, as shown in Figure 16-3.

Figure 16-3:
The
progress
loader in the
status bar.

As a developer, you have programmatic access to provide custom expanded views. The expanded view is the view that is present when the user taps the status bar icon. Fore more information on custom views inside the expanded view, see the Android documentation here: `http://d.android.com/guide/topics/ui/notifiers/notifications.html`

Using the Notification Manager to Create Your First Notification

The notification manager allows you to interface with Android's notification mechanism. Notifications appear in the status bar at the bottom of the device screen. Working with the `NotificationManager` is as simple as asking the current context for it. If you are within an activity, the code is as follows:

```
NotificationManager mgr = (NotificationManager)getSystemService(NOTIFICATION_
        SERVICE);
```

This line of code obtains the `NotificationManager` object from the `get-SystemService()` call.

The Task Reminder application needs a way to notify the user that a task needs attention. A notification would show up when the alarm goes off for that particular task. To set this notification in the status bar, you need to use the `NotificationManager` object.

In the `doReminderWork()` method of the `ReminderService` class, type the code as shown in Listing 16-1.

Listing 16-1: Implementation of doReminderWork()

```
Long rowId = intent.getExtras().getLong(RemindersDbAdapter.KEY_ROWID);    →1

NotificationManager mgr =
            (NotificationManager)getSystemService(NOTIFICATION_SERVICE);  →4

Intent notificationIntent = new Intent(this, ReminderEditActivity.class); →6
notificationIntent.putExtra(RemindersDbAdapter.KEY_ROWID, rowId);         →7

PendingIntent pi = PendingIntent.getActivity(this, 0, notificationIntent,
            PendingIntent.FLAG_ONE_SHOT);                                 →9

Notification note=new Notification(android.R.drawable.stat_sys_warning,
            getString(R.string.notify_new_task_message),
                System.currentTimeMillis());                              →13

note.setLatestEventInfo(this, getString(R.string.notifiy_new_task_title),
            getString(R.string.notify_new_task_message), pi);            →16

note.defaults |= Notification.DEFAULT_SOUND;                             →18
note.flags |= Notification.FLAG_AUTO_CANCEL;                             →19

// An issue could occur if user ever enters over 2,147,483,647 tasks. (Max int
            value).
// I highly doubt this will ever happen. But is good to note.
int id = (int)((long)rowId);                                            →23
mgr.notify(id, note);                                                   →24
```

The various lines of Listing 16-1 are explained as follows:

→1   The intent that started the `ReminderService` contains the row ID of the task that you're currently working with. You need this ID because you will set this as part of the `PendingIntent` for the status. When the notification is selected from the status bar, you want the `ReminderEditActivity` to start with the row ID as part of the pending intent. That way, the `ReminderEditActivity` will open, read the data about that particular row ID, and display it to the user.

→**4**   Get an instance of the `NotificationManager`.

→**6**   In this line you build a new intent and set the class to `ReminderEditActivity`. This is the activity that you want to start when the user selects the notification.

→**7**   Put the row ID into the intent.

→**9**   Set up a pending intent to be used by the notification system. Because the notification system runs in another process, a `PendingIntent` is required. The `FLAG_ONE_SHOT` flag is used to indicate that this pending intent can only be used once.

→**13**   This line builds the `Notification` that shows up in the status bar. The `Notification` class accepts the following parameters:

- `icon`: `android.R.drawable.stat_sys_warning`: The resource ID of the icon to place in the status bar. This icon is a small triangle with an exclamation point in the middle. Because this is a built-in Android icon, you don't have to worry about providing small-, medium-, or high-density graphics — they are already built into the platform.

- `tickerText`: `getString(R.string.notify_new_task_message)`: The text that flows by when the notification first activates.

- `when`: `System.currentTimeMillis()`: The time to display in the time field of the notification.

→**16**   This line sets the content of the expanded view with that standard Latest Event layout provided by Android. If you wanted to, you could provide a custom XML layout to display. However, in this instance, you simply provide the stock notification view. The `set-LatestEventInfo()` method accepts the following parameters:

- `context`: `this`: The context to associate with the event info

- `contentTitle`: `getString(R.string.notifiy_new_task_title)`: The title that goes into the expanded view

- `contextText`: `getString(R.string.notify_new_task_message)`: The text that goes into the expanded view

- `contentIntent`: `pi`: The intent to launch when the expanded view is selected

→**18**   A bitwise-ored in setting the `Notification` object to include sound during the notification process. This forces the default notification sound to be played if the user has the notification volume on.

→**19**   A bitwise-ored in setting the `Notification` object flag's property that cancels the notification after it is selected by the user.

→23 Casting the ID to an integer. The ID stored in the SQLite database is long; however, here you cast it to an integer, which admittedly loses a bit of precision. However, I highly doubt that this application would ever need to set up more than 2,147,483,647 tasks (which is the maximum number that an integer can store in Java). Therefore, this casting should be okay. The casting to an integer is necessary because the code on line 24 only accepts an integer as the ID for the notification.

→24 Raises the notification to the status bar. The `notify()` call accepts two parameters:

- `id: id`: An ID that is unique within your application.

- `Notification: note`: A `Notification` object that describes how to notify the user.

The code in Listing 16-1 allows the following workflow to occur:

1. The user is active in another application, such as e-mail.

2. A task is due and therefore the alarm fires. The notification is created in the status bar.

3. The user can elect either to slide down the status bar and select the notification or to ignore it for now.

   If the user chooses to slide open the status bar and select an item, the pending intent within the notification will be activated. This in turn causes the `ReminderEditActivity` to open with the given row ID of the task.

4. The notification is removed from the status bar.

5. The task information is retrieved from the database and displayed on the form in the `ReminderEditActivity`.

You may notice that you need to add the following two string resources:

✔ `notify_new_task_message`: I have set the value of this to "A task needs to be reviewed!" This message is used as the message in the expanded view and is used as the ticker text when the notification first arrives.

✔ `notify_new_task_title`: I have set the value of this to "Task Reminder." This message is used as the title for the expanded view.

Updating a Notification

At some time, you may need to update the view of your notification. Consider the following situation: You have some code that runs in the background to see whether the tasks have been reviewed. This code checks to see whether any notifications are overdue. You decide that after the two-hour mark passes, you want to change the icon of the notification to a red-colored exclamation point and flash the LED quickly with a red color. Thankfully, updating the notification is a fairly simple process.

If you call one of the `notify()` methods with an ID that is currently active in the status bar, and with a new set of notification parameters, the notification is updated in the status bar. Therefore, to update your notification, simply create a new `Notification` object with the red icon as a parameter and call `notify()`.

Clearing a Notification

Users are highly unpredictable — they could be anyone, anywhere in the world! They could be first-time users, advanced power users, and so on. Each user uses the device in his or her own special way, including ways you may not predict. Case in point: What happens if your user sees a notification and decides to open the app the manual/long way — via the app launcher?

If the user decides to open your application via the app launcher while a notification is active, your notification will persist. Even if the user looks at the task at hand, the notification will still persist on the status bar. Although this is not a big deal, your application should be able to recognize the state of the application and should take the appropriate measures to cancel any existing notifications that might be present for the given task. However, if the user opens your app and reviews a different task that does not have an active notification, you should not clear any notifications. Only clear the notification for tasks that the user is reviewing.

With the `cancel()` method, the `NotificationManager` makes it simple to cancel an existing notification. The `cancel()` method accepts one parameter — the ID of the notification. Remember how you used the ID of the task as the ID for the note? This is why you did that. The ID of the task is unique to the Task Reminder application. By doing this, you can easily open a task and cancel any existing notification by calling the `cancel()` method with the ID of the task.

At some point, you might also need to clear all previously shown notifications. To do this, simply call the `cancelAll()` method on the `NotificationManager`.

Chapter 17

Working with Android's Preference Framework

. . . .

In This Chapter

▶ Seeing how preferences work in Android

▶ Building a preference screen

▶ Working with preferences programmatically

. . . .

I consider myself a power user of computer software, and I'm sure that you're a power user as well. I know that most programs can be configured to suit my needs (for the most part), and I usually go out of my way to find the settings or preferences to set up my favorite configuration for a given program. Allowing your users to do the same in your Android application gives your application an advantage in regard to usability. Thankfully, creating and providing a mechanism to edit preferences in Android are fairly easy processes.

Out of the box, Android provides a robust preference framework that allows you to declaratively as well as programmatically define preferences for your application. Android stores preferences as persistent key-value pairs of primitive data types for you. You are not required to store the values in a file, database, or any other mechanism. The Android preference framework takes the values you provide and commits them to internal storage on behalf of your application. You can use the preference framework to store booleans, floats, ints, longs, and strings. The data persists across user sessions as well — meaning that if the user closes the app and reopens it later, the preferences are saved and can be used again. This is true even if your application is killed.

In this chapter, I delve into the Android preference framework and describe how to incorporate it into your applications. I demonstrate how to use the built-in `PreferenceActivity` to create and edit preferences. I also demonstrate how to read and write preferences from code within your application. By the end of the chapter, you will have fully integrated preferences into the Task Reminder application.

Understanding Android's Preference Framework

One of the great things about the Android preference framework is how simple it is to develop a screen that allows the user to modify preferences. Most of the heavy lifting is done for you by Android, because developing a preference screen is as simple as defining a preference screen in XML that is located in the res/xml folder of your project. Although these XML files are not the same as layout files, they are specific XML definitions that define screens, categories, and actual preferences. Common preferences that are built into the framework include the following:

✔ EditTextPreference: A preference that can store plain text as a string

✔ CheckBoxPreference: A preference that can store a boolean value

✔ ListPreference: A preference that allows the user to select a preferred item from a list of items in the dialog box

If the built-in preferences do not suit your needs, you can create your own preference by deriving from the base Preference class or from DialogPreference — the base class for dialog box–based preferences. When clicked, these preferences open a dialog box showing the actual preference controls. Examples of built-in DialogPreferences are EditTextPreference and ListPreference.

Android also provides a PreferenceActivity from which you can derive and load the preference screens in the same manner that you would load a layout for a basic Activity class. The PreferenceActivity base class allows you to tap into the PreferenceActivity events and perform some advanced work, such as setting an EditTextPreference to accept only.

Understanding the PreferenceActivity Class

The responsibility of the PreferenceActivity class is to show a hierarchy of Preference objects as lists, possibly spanning multiple screens, as shown in Figure 17-1.

When preferences are edited, they are stored using an instance of the SharedPreferences class, which is an interface for accessing and modifying preference data returned by getSharedPreferences() from any Context object.

A preference screen with various preferences loaded

A `PreferenceActivity` is a base class that is very similar to the `Activity` base class. However, the `PreferenceActivity` behaves a bit differently. One of the most important features that the `PreferenceActivity` handles is the display of preferences in the visual style that resembles the system preferences. This gives your application a feel consistent with other Android user interface components. You use the `PreferenceActivity` when dealing with preference screens in your Android applications.

Persisting preference values

Because the Android framework stores preferences in the `SharedPreferences`, which automatically stores the preference data in internal storage, creating a preference is easy. When a user edits a preference, the value is automatically saved for you — that's right, you don't have to do any persisting yourself!

I'm sure this sounds like a little bit of black magic, but I assure you it's not! In Figure 17-2, I am editing an `EditTextPreference` that will be used in the Task Reminder application. After I select OK, Android takes the value I provided and persists it to `SharedPreferences` — I don't need to do anything else. Android does all the heavy lifting in regard to persisting the preference values.

Figure 17-2:
Setting a
preference.

Laying out preferences

Working with layouts in Android can sometimes be, well, a painstaking process of alignment, gravity, and so on. Building layouts is almost like building
a website with various tables all over the place. Sometimes it's easy; sometimes it's not. Thankfully, laying out Android preferences is much simpler
than defining a layout for your application screen.

Android preference screens are broken into the following categories:

✔ PreferenceScreen: Represents a top-level preference that is the root of
a preference hierarchy. You can use a PreferenceScreen in two places:

• In a PreferenceActivity: The PreferenceScreen is not
shown because it only shows the containing preferences within the
PreferenceScreen definition.

• In another preference hierarchy: When present in another hierarchy,
the PreferenceScreen serves as a gateway to another screen of
preferences. Think of this as nesting PreferenceScreen declarations inside other PreferenceScreen declarations. Although
this might seem confusing, think of this as XML. In XML, any element
you declare can contain itself as its own parent element. At
that point, you're nesting the elements. The same goes for the
PreferenceScreen. By nesting them, you are informing Android
that it should show a new screen when selected.

✔ PreferenceCategory: This preference is used to group preference objects and provide a title above the group that describes the category.

✔ Preference: A preference that is shown on the screen. This preference could be any of the common preferences or a custom one that you define.

By laying out a combination of the PreferenceScreen, PreferenceCategory, and Preference in XML, you can easily create a preference screen that looks similar to Figure 17-1.

Creating Your First Preference Screen

Creating preferences using the PreferenceActivity and a preference XML file is a fairly straightforward process. The first thing you do is create the preference XML file, which defines the layout of the preferences and the string resource values that show up on the screen. These string resources are presented as TextViews on the screen to help the user determine what the preference is for.

The PreferenceScreen I am building is for the Task Reminder application. I want to be able to give my users the chance to set the default time for a reminder (in minutes) and a default title for a new task. As the application stands right now, the default title is empty and the default reminder time is set to the current time. These preferences allow the user to save a couple of steps while building new tasks. For example, if the user normally builds tasks with a reminder time of 60 minutes from now, the user can now set that in the preferences. This new value becomes the value of the reminder time when the user creates a new task.

Building the preferences file

To build your first preference screen, you need to create a res/xml folder in your project. Inside the res/xml folder, create an XML file — I'm naming mine task_preferences.xml. Listing 17-1 outlines what should be in the file.

Listing 17-1:   The task_preferences.xml File

```
<?xml version="1.0" encoding="utf-8"?>
<PreferenceScreen                                                         →2
    xmlns:android="http://schemas.android.com/apk/res/android">
        <PreferenceCategory                                               →4
            android:key="@string/pref_category_task_defaults_key"         →5
            android:title="@string/pref_category_task_defaults_title">    →6
```

(continued)

Listing 17-1 *(continued)*

```
        <EditTextPreference                                              →7
            android:key="@string/pref_task_title_key"                    →8
            android:dialogTitle="@string/pref_task_title_dialog_title"   →9
            android:dialogMessage="@string/pref_task_title_message"      →10
            android:summary="@string/pref_task_title_summary"            →11
            android:title="@string/pref_task_title_title" />             →12
    </PreferenceCategory>
    <PreferenceCategory                                                  →14
        android:key="@string/pref_category_datetime_key"                 →15
        android:title="@string/pref_category_datetime_title">            →16
        <EditTextPreference                                              →17
            android:key="@string/pref_default_time_from_now_key"         →18
    android:dialogTitle="@string/pref_default_time_from_now_dialog_title" →19
    android:dialogMessage="@string/pref_default_time_from_now_message"   →20
    android:summary="@string/pref_default_time_from_now_summary"         →21
    android:title="@string/pref_default_time_from_now_title" />          →22
    </PreferenceCategory>
</PreferenceScreen>
```

Quite a few string resources are introduced in Listing 17-1. They are listed in Listing 17-2. Each numbered line of code is explained as follows:

→**2**   This line is the root-level PreferenceScreen. It is the container for the screen itself. All other preferences live below this declaration.

→**4**   This line is a PreferenceCategory that defines the category for task defaults, such as title or body. As you may have noticed, on line 13, you declare another PreferenceCategory for the default task time. Normally you would have placed these two items into the same category, but in this instance I asked you to separate them to demonstrate how to use multiple PreferenceCategory elements on one screen.

→**5**   This line defines the key that is used to store and retrieve the preference from the SharedPreferences. This key must be unique.

→**6**   This line defines the category title.

→**7**   This line contains the definition of the EditTextPreference, which is responsible for storing the preference for the default title of a task.

→**8**   This line contains the key for the default title text EditTextPreference.

→**9**   The EditTextPreference is a child class of DialogPreference, which means that when you select the preference, you will receive a dialog box similar to the one shown in Figure 17-2. This line of code defines the title for that dialog box.

→**10**   This line defines the message that appears in the dialog box.

→**11**   This line defines the summary text that is present on the preferences screen, as shown in Figure 17-1.

→**12**   This line defines the title of the preference on the preference screen.

→**14**   This line defines the `PreferenceCategory` for the default task time.

→**15**   This line defines the category key.

→**16**   This line defines the title of the category.

→**17**   This line is the start of the definition of the `EditTextPreference`, which stores the default time in minutes (digits) that the task reminder time will default to from the current time.

→**18**   This line defines the key for the default task time preference.

→**19**   This line defines the title of the dialog box that presents when the preference is selected.

→**20**   This line defines the message that will be present in the dialog box.

→**21**   This line defines the summary of the preference that is present on the main preference screen, as shown in Figure 17-1.

→**22**   This line defines the title of the preference on the preference screen.

Adding string resources

For your application to compile, you need the string resources for the preferences. In your `res/values/strings.xml` file, add the following values:

```xml
<!-- Preferences -->
<string name="pref_category_task_defaults_key">task_default_category</string>
<string name="pref_category_task_defaults_title">Task Title Default</string>
<string name="pref_task_title_key">default_reminder_title</string>
<string name="pref_task_title_dialog_title">Default Reminder Title</string>
<string name="pref_task_title_message">The default title for a reminder.</string>
<string name="pref_task_title_summary">Default title for reminders.</string>
<string name="pref_task_title_title">Default Reminder Title</string>
<string name="pref_category_datetime_key">date_time_default_category</string>
<string name="pref_category_datetime_title">Date Time Defaults</string>
<string name="pref_default_time_from_now_key">time_from_now_default</string>
<string name="pref_default_time_from_now_dialog_title">Time From Now</string>
<string name="pref_default_time_from_now_message">The default time from now (in minutes) that a new reminder should be set to.</string>
<string name="pref_default_time_from_now_summary">Sets the default time for a reminder.</string>
<string name="pref_default_time_from_now_title">Default Reminder Time</string>
```

You should now be able to compile your application.

Defining a preference screen was fairly simple — provide the values to the attributes needed and you're done. Although the preference screen may be defined in XML, simply defining it in XML does not mean that it will show up on the screen. To get your preference screen to display on the screen, you need to create a PreferenceActivity.

Working with the PreferenceActivity Class

The PreferenceActivity shows a hierarchy of preferences on the screen according to a preferences file defined in XML — such as the one you just created. The preferences can span multiple screens (if multiple PreferenceScreen objects are present and nested). These preferences automatically save to SharedPreferences. As an added bonus, the preferences shown automatically follow the visual style of the system preferences, which allows your application to have a consistent user experience in conjunction with the default Android platform.

To inflate and display the PreferenceScreen you just built, add an activity that derives from PreferenceActivity to your application. I am going to name mine TaskPreferences. Please add this file to the src/ directory of your project. The code for this file is shown in Listing 17-2.

Listing 17-2: The TaskPreferences File

```
public class TaskPreferences extends PreferenceActivity {          →1
    @Override
    protected void onCreate(Bundle savedInstanceState) {
        super.onCreate(savedInstanceState);
        addPreferencesFromResource(R.xml.task_preferences);        →5

        EditTextPreference timeDefault = (EditTextPreference)
    findPreference(getString(R.string.pref_default_time_from_now_key));  →8
  timeDefault.getEditText().setKeyListener(DigitsKeyListener.getInstance());  →9
    }
}
```

Yes, that's it! That's all the code that is needed to display, edit, and persist preferences in Android! Each numbered line of code is explained as follows:

→1   The TaskPreferences class file is defined by inheriting from the PreferenceActivity base class.

→5   The call to addPreferencesFromResource() method is provided with the resource ID of the task_preferences.xml file stored in the res/xml directory.

→8   Here you retrieve the `EditTextPreference` for the default task reminder time by calling the `findPreference()` method and providing it with the key that was defined in the `task_preferences.xml` file.

→9   On this line, you obtain the `EditText` object, which is a child of the `EditTextPreference`, using the `getEditText()` method. From this object, you set the key listener, which is responsible for listening to key-press events. You set the key listener through the `setKeyListener()` method, and by providing it with an instance of `DigitsKeyListener`, the `EditTextPreference` allows only digits to be typed into the `EditTextPreference` for the default reminder time. This is because you do not want users to type string values such as `foo` or `bar` into the field — these are not valid integer values. Using the `DigitsKeyListener` ensures that only integer values get passed into the preferences.

At this point, the activity is ready to be used. This `PreferenceActivity` allows users to edit and save their preferences. As you can see, this implementation required very little code. The next step is getting the preference screen to show up by adding a menu item for it.

Don't forget! You also need to add your new `PreferenceActivity` to your `AndroidManifest.xml` file with the following line of code:

```
<activity android:name=".TaskPreferences" android:label="@string/app_name" />
```

Opening the PreferenceActivity class

To open this new activity, you need to add a menu item to the `ReminderListActivity`. To add a new menu item, you need to add a new menu definition to the `list_menu.xml` file located in the `res/menu` directory. Updating this file updates the menu on the `ReminderListActivity`. The updated `list_menu.xml` file is shown as follows with the new entry bolded:

```
<?xml version="1.0" encoding="utf-8"?>
<menu
    xmlns:android="http://schemas.android.com/apk/res/android">
    <item android:id="@+id/menu_insert"
        android:icon="@android:drawable/ic_menu_add"
        android:title="@string/menu_insert" />
    <item android:id="@+id/menu_settings"
        android:icon="@android:drawable/ic_menu_preferences"
        android:title="@string/menu_settings" />
</menu>
```

The last item adds a menu item for settings, which uses the built-in Android settings icon and a string resource called `menu_settings`. You need to add a new string resource called `menu_settings` with a value of `Settings` in your string resources.

Handling menu selections

Now that you have your menu updated, you need to be able to respond to when the menu item is tapped. To do so, you need to add code to the `onMenuItemSelected()` method in the `ReminderListActivity`. The code to handle the settings menu selection is bolded:

```
@Override
public boolean onMenuItemSelected(int featureId, MenuItem item) {
        switch(item.getItemId()) {
            case R.id.menu_insert:
                createReminder();
                return true;
            case R.id.menu_settings:
                Intent i = new Intent(this, TaskPreferences.class);
                startActivity(i);
                return true;
            }
        return super.onMenuItemSelected(featureId, item);
}
```

The bolded code here simply creates a new `Intent` object with a destination class of `TaskPreferences`. When the user selects the Settings menu item, he is now shown the preferences screen, where he can edit the preferences. If you start the app and select Settings, you should see a preferences screen similar to the one in Figure 17-3.

Figure 17-3:
The pref-
erences
screen.

Working with Preferences in Your Activities at Run Time

Although setting preferences in a `PreferenceActivity` is useful, in the end, it provides no actual value unless you can read the preferences from the `SharedPreferences` object at run time and use them in your application. Thankfully, Android makes this process simple.

In the Task Reminder application, you need to read these values in the `ReminderEditActivity` to set the default values when a user creates a new task. Because the preferences are stored in `SharedPreferences`, you can access the preferences across various activities in your application.

Retrieving preference values

Open the `ReminderEditActivity` and navigate to the `populateFields()` method. This method determines whether the task is an existing task or a new task. If the task is new, you will pull the default values from `SharedPreferences` and load them into the activity for the user. If for some reason the user has never set the preferences, these default values will be empty strings, and you will ignore them. In short, you will only use the preferences if the user has set them.

To retrieve the preference values, you need to use the `SharedPreferences` object, as shown in Listing 17-3. In the `populateFields()` method, add the bolded code as shown in Listing 17-3.

Listing 17-3:   Retrieving Values from SharedPreferences

```
private void populateFields()  {
    if (mRowId != null) {
        Cursor reminder = mDbHelper.fetchReminder(mRowId);
        startManagingCursor(reminder);
        mTitleText.setText(reminder.getString(
        reminder.getColumnIndexOrThrow(RemindersDbAdapter.KEY_TITLE)));
        mBodyText.setText(reminder.getString(
        reminder.getColumnIndexOrThrow(RemindersDbAdapter.KEY_BODY)));

        SimpleDateFormat dateTimeFormat = new
            SimpleDateFormat(DATE_TIME_FORMAT);
        Date date = null;
        try {
            String dateString =
                reminder.getString(reminder.getColumnIndexOrThrow(
                    RemindersDbAdapter.KEY_DATE_TIME));
            date = dateTimeFormat.parse(dateString);
```

(continued)

Listing 17-3 *(continued)*

```
            mCalendar.setTime(date);
        } catch (IllegalArgumentException e) {
            e.printStackTrace();
        } catch (ParseException e) {
            e.printStackTrace();
        }
    } else {                                                              →25
    SharedPreferences prefs =
        PreferenceManager.getDefaultSharedPreferences(this);             →27
        String defaultTitleKey = getString(R.string.pref_task_title_key); →28
        String defaultTimeKey =
                getString(R.string.pref_default_time_from_now_key);       →30

        String defaultTitle = prefs.getString(defaultTitleKey, "");      →32
        String defaultTime = prefs.getString(defaultTimeKey, "");        →33
        if("".equals(defaultTitle) == false)
            mTitleText.setText(defaultTitle);                            →35
        if("".equals(defaultTime) == false)
            mCalendar.add(Calendar.MINUTE, Integer.parseInt(defaultTime)); →37
    }

    updateDateButtonText();
    updateTimeButtonText();                                              →41
}
```

Each new line of code is explained as follows:

→**25**   The `else` statement to handle the logic for a new task.

→**27**   This line retrieves the `SharedPreferences` object from
           the static `getDefaultSharedPreferences()` call on the
           `PreferenceManager` object.

→**28**   On this line, you retrieve the key value for the default title prefer-
           ence from the string resources. This is the same key that is used
           in Listing 17-1 to define the preference.

→**30**   On this line, you retrieve the key value for the default time offset,
           in minutes, from the preferences (a different key but the same pro-
           cess as line 28).

→**32**   On this line, you retrieve the default title value from the preferences
           with a call to `getString()` on the `SharedPreferences` object.
           The first parameter is the key for the preference, and the second
           parameter is the default value if the preference does not exist (or
           has not been set). In this instance, you request that the default
           value be `" "` (an empty string) if the preference does not exist.

→**33**   On this line, you retrieve the default time value from the preferences, using the same method as described on line 32 with a different key.

→**35**   On this line, you set the text value of the `EditText` view — which is the title of the task. You set this value only if the preference was not equal to an empty string.

→**37**   On this line, you increment time on the local `Calendar` object by calling the `add()` method with the parameter of `Calendar.MINUTE` (only if the value from the preferences was not equal to an empty string). The `Calendar.MINUTE` constant informs the `Calendar` object that the next parameter should be treated as minutes and the value should get added to the calendar's minute field. If the minutes force the calendar into a new hour or day, the calendar object updates the other fields for you. For example, if the calendar was originally set to 2010-12-31 11:45 p.m. and you added 60 minutes to the calendar, the new value of the calendar would be 2011-01-01 12:45 a.m. Because the `EditTextPreference` stores all values as strings, you cast the string minute value to an integer with the `Integer.parseInt()` method. By adding time to the local `Calendar` object, the time picker and button text associated with opening the time picker are updated as well.

→**41**   On this line, you update the time button text to reflect the time that was added to the existing local `Calendar` object.

When you start the application, you can now set the preferences and see them reflected when you choose to add a new task to the list. Try clearing the preferences and then creating a new task. Notice that the defaults no longer apply. Wow, that was easy!

Setting preference values

At times, you may need to update preference values through code instead of preference screens. (This approach is not necessary for the Task Reminder application, however.) Consider the following case: You develop a help desk ticket-system application that requires the user to enter his or her current department on each ticket. You have a preference for the default department, but the user never uses the preferences screen. Instead, she repeatedly enters the department for each ticket by hand into your application. Through the logic you defined, your app determines that the user is entering the same department (the Accounting department, say) over and over again for each help desk ticket, so it prompts the user to specify whether she would like to set the default department to Accounting. If the user chooses Yes, the app programmatically updates the preferences for her. I show you how to do that now.

To edit preferences programmatically, you need an instance of `SharedPreferences`. You can obtain that through the `PreferenceManager`, as shown in Listing 17-4. After you obtain an instance of `SharedPreferences`, you can edit various preferences by obtaining an instance of the preference `Editor` object. After the preferences are edited, you need to commit them. This process is also demonstrated in Listing 17-4.

Listing 17-4: Programmatically Editing Preferences

```
SharedPreferences prefs =                                             →2
    PreferenceManager.getDefaultSharedPreferences(this);
Editor editor = prefs.edit();                                         →3
editor.putString("default_department", "Accounting");                →4
editor.commit();                                                      →5
```

Each numbered line of code is explained as follows:

→2   An instance of `SharedPreferences` is retrieved from the `PreferenceManager`.

→3   An instance of the preferences `Editor` object is obtained by calling the `edit()` method on the `SharedPreferences` object.

→4   On this line, you edit a preference with the key value of `default_department` by calling the `putString()` method on the `Editor` object. You set the value to "Accounting". Normally, the key value would be retrieved from the string resources and the value of the string would be retrieved through your program or user input. The code snippet is kept simple for brevity.

→5   After changes are made to any of the preferences, you must call the `commit()` method on the `Editor` object to persist them to `SharedPreferences`. The commit call automatically replaces any value that is currently stored in `SharedPreferences` with the key given in the `putString()` call.

If you do not call `commit()` on the `Editor` object, your changes will not persist and your application will not function as you expect.

Part IV
The Part of Tens

The 5th Wave · By Rich Tennant

"Has the old media been delivered yet?"

In this part . . .

Part IV consists of some of the best secret-sauce-covered Android nuggets — the sort of information you can acquire only after having been in the development trenches for quite some time. First, I list some of the best sample applications that can help springboard you on your way to creating the next hit application. These applications include database-oriented apps, interactive games, and applications that interact with third-party Web application programming interfaces (APIs).

I close Part IV with a list of professional tools and libraries that can help streamline and improve the productivity of your application development process and make your life as a developer much easier.

Chapter 18

Ten Great Free Sample Applications and SDKs (with Code!)

During your career as an Android developer, you may run into various roadblocks — including code-based roadblocks. Say you're looking to communicate with a third-party API that returns JSON, or you're trying to perform collision detection in a game, but you're not sure how to go about it. When you run into these sorts of problems, one solution is to search the web for sample code. Chances are, someone out there has already written the code you're after! You can then review that code, alter it as needed, and continue with development.

Sample code is great, but it's just that — sample code. It's not production ready, and you usually can't just plug it into your application without first making some adjustments. However, sample code has a valuable side effect: It is a learning enhancer. A good way to find out how to program for Android is to analyze sample code. Sure, sample code — such as the API demos I mention in Chapter 4 — comes with the Android SDK. But a plethora of real-world application code is freely available on the web! You can find plenty of good-quality, open-source applications on the Internet, which can serve as great learning tools.

Telling you to find them yourself would be rather rude, now wouldn't it? To help speed up your learning process, this chapter presents ten really cool open-source applications and samples for you to explore. Most of the source code examples that follow are real-world Android applications you can install from the Android Market. I advise you to download these applications on your device and interact with them. Then, crack open the source code to see how the gears work!

TekPub Video Player

TekPub.com is a technical video training site. I created their Android video series. I built a video-player application that plays the videos that users can access. The application uses RoboGuice, gson (Google library for working with JSON objects), HTTP communication, the native video player, the Account Manager, and many other features. The code is completely open source and free to use. Find out how to use the Account Manager in Android and how to communicate asynchronously with an HTTP web API as well. Download: `https://github.com/tekpub/IntroductionToAndroid`.

Last.fm App Suite

Are you the next up-and-coming Internet radio sensation? If so, you might want to find out how to stream music by using the Last.fm API as an example. To run and test the app, you need a Last.fm API key that you can obtain by visiting this URL: `www.last.fm/api/account`. You also need a paid account to stream music; however, a paid account isn't necessary to review the code. You don't need to apply for a key or pay for an account if you simply want to review the source code. This example can help you understand the fundamentals of streaming music from a remote location. Source code: `http://github.com/mxcl/lastfm-android`.

WordPress for Android

You can't visit the Internet without running into a site running WordPress. From prolific bloggers to major corporations, they all use it. The WordPress for Android application is built by the actual WordPress team for managing a wordpress installation. Learn how to make XMLRPC calls in addition to other cool tidbits by reading through the source code. `http://android.svn.wordpress.org/trunk/`

LOLCat

This is a great example if you are interested in image manipulation with Android. You find out how to take a picture using the device's camera, add captions to it, and then save the resulting file on the SD Card. You also discover how to create various intents, which allow you to send the image as an MMS (multimedia messaging service) image or as an e-mail attachment. Source code: `http://code.google.com/p/apps-for-android`.

Amazed

Amazed is a fun game that can demonstrate the use of the device's built-in accelerometer to control a 2D marble through various obstacles inside increasingly difficult maze levels. If you are interested in accelerometer-based applications, reviewing this application source code can help you immensely. Not only does the application show you how to use the accelerometer, it also demonstrates other game development fundamentals such as collision detection and the game loop principle. Source code: `http://code.google.com/p/apps-for-android`.

APIDemos

The Android SDK provides various sample applications, one of which is the API Demos application. This application demonstrates how to use the various Android APIs through small, digestible, working examples. You find tons of simple straight-to-the-point examples in the API Demos source code. Perhaps you're interested in incorporating animation into your project, or you want to play an audio file inside your app — that's easy because the API Demos provide examples of both! If you have a lot of ideas but not a lot of time, you should definitely check out these cool examples. I recommend installing this demo app on your device and playing with each of the numerous examples to see exactly what they can do. Source code: In your Android SDK, in the `samples` folder.

Hubroid

Git is a popular open-source Distributed Version Control System (DVCS), and actually, all the code and documents written for this book were stored in various Git repositories during the writing! Hubroid is a GitHub.com-based application for Android that allows you to view all your favorite Git repositories located on GitHub.com from the palm of your hand. Hubroid demonstrates how to use the GitHub API. If you want to work with the GitHub API, this code is a great resource on how to "Git 'er done." Source code: `http://github.com/eddieringle/hubroid`.

Facebook SDK for Android

Are you feeling ambitious? If so, you might want to tackle the task of creating the next best Facebook application, but maybe you don't know where to begin. The Facebook Android SDK enables you to integrate Facebook functionality into your application easily. You can use it to authorize users, make API requests, and much more! Integrate all the Facebook goodness without breaking a sweat. Source code: `http://github.com/facebook/facebook-android-sdk`.

Replica Island

Perhaps you want to make a side-scrolling game but have no clue how to get started. Well, it's your lucky day because Replica Island is a very cool side-scrolling game that features none other than the little green robot that we know and love — the Android. Not only is it a popular free game on the Android Market, it's also completely open source and a great learning tool for game developers! This truly is a great example of a 2D game for the Android platform. Source code: `http://code.google.com/p/replicaisland`.

Notepad Tutorial

If you're interested in finding out how to use the basics of SQLite without all the other fluff of services, background tasks, and so on, this app is for you. Although simple in its execution and usage, the source code and tutorial that go along with it help you understand the basics of SQLite. Source code and tutorial: `http://d.android.com/guide/tutorials/notepad/index.html`.

Chapter 19

Ten Tools That Make Your Developing Life Easier

*A*s a developer, you inherently will build tools to help yourself become more productive. I have created various helper methods to assist in asynchronous communication, XML and JSON parsing, date and time utilities, and much more. Before you write a ton of helper classes or frameworks to handle items for you, I advise you to look on the Internet for tools that already exist. In this chapter, I compile a list of ten tools and utilities that can make your developer life much easier by increasing your productivity and ensuring that your app is up to snuff.

RoboGuice

No, it's not the latest and greatest energy drink marketed to developers. RoboGuice is a framework that uses Google's Guice library to make dependency injection a breeze. Dependency injection handles initializing your variables at the right time so that you don't have to. This really cuts down the amount of code you have to write overall, and it makes maintaining your application a breeze in the future. Source code: `http://code.google.com/p/roboguice`.

IntelliJ IDEA

Some developers, myself included, are not huge fans of Eclipse. Although the integrated development environment works, it's not optimal for development, in my opinion. I'm a huge fan of Jet Brain's products and use them all over my development arena. IntelliJ IDEA is a Java IDE that allows you to build Android (and other Java) applications. The community version is free, so download it and build your next Android app with it! Note that the Google Android team only develops the ADT plug-in for Eclipse, therefore any updates that need to be applied to IntelliJ IDEA will have to wait until the Jet Brains team gets them implemented. Thankfully, the team is quite quick at implementing changes. Download: `www.jetbrains.com/idea/`.

TeamCity

This is another phenomenal tool from Jet Brains for helping with application development. TeamCity is a continuous integration server that monitors your source control repository (Git, Mercurial, Subversion, and many more) and performs actions upon check-in of new code. I currently have a TeamCity set up on a server, monitoring my android projects. When a check-in occurs (initiated by me or a teammate), TeamCity builds the Android project and provides me with a built APK (known as an artifact). I can then allow a quality assurance tester to log into the server and download a new build of my applications for testing. TeamCity has a free Professional Version (which I run) and can manage up to 20 different builds. Download: `www.jetbrains.com/teamcity/`.

Git

Git is a super-fast, free, and open-source-distributed version-control system. Git manages repositories quickly and efficiently, making it painless to back up your work in a cinch. Don't let a system crash ruin your day by not having a version-control system for your next spectacular app! Git makes working with branching very simple and effective, and it integrates into your workflow very easily. Eclipse plug-ins exist to help manage your Git repository from within the Eclipse IDE. Although Git is distributed, you will most likely want to have a remote location where the Git repository is stored. You can obtain a free private Git repository from Projectlocker.com or Unfuddle.com. If your code is open source, you can create free repositories on Github.com. You can find more information at `http://git-scm.com`.

Gson

Gson is a library that is provided by Google to help with the serialization and deserialization of Json to and from POJOs (Plain Old Java Objects). This library allows you to model a Java object with the same names as a JSON document and then have Gson return a Java object based upon the JSON. No more manually parsing JSON! Gson also works the other way around — it will take a Java object and convert it to JSON for you. Talk about productive! Download: `http://code.google.com/p/google-gson/`.

droid-fu

Droid-fu is an open-source library with a handful of methods that can karate-chop your development time drastically. Droid-fu is comprised of utility classes that do all the mundane heavy lifting for you, such as handling asynchronous background requests, retrieving images from the web, and, most amazingly, enhancing the application life cycle. Never worry about state changes because droid-fu handles all of it and much more. Don't just sit there. Start earning your black belt in droid-fu today! Source code: `http://github.com/kaeppler/droid-fu`.

Draw 9-patch

Draw 9-patch is a utility that enables you to easily create scalable images for Android. While Draw 9-patch images were not discussed in this book, you can find more detail here: `http://d.android.com/guide/developing/tools/draw9patch.html`.

You use this utility to embed instructions in your image to tell the OS where to stretch your images so that they display as crisp and clean as possible regardless of the size or resolution of the device screen.

Hierarchy Viewer

Working with various views inside your layout file to create a UI isn't always as straightforward as you want it to be. The Hierarchy Viewer, located in the Android SDK `tools` directory, lets you see exactly how your widgets are laid out on the screen in a graphical representation. This format allows you to clearly see each widget's boundaries so that you can determine what's going on inside your layout. This is the ultimate tool to make pixel-perfect UIs. The Hierarchy Viewer also lets you magnify the display in the pixel-perfect view to make sure that your images and UIs will display flawlessly on all screen sizes and densities. You can read all about it at `http://developer.android.com/guide/developing/tools/hierarchy-viewer.html`.

UI/Application Exerciser Monkey

Don't worry — this monkey doesn't need to be fed bananas to remain happy! You use Exerciser Monkey to stress-test your application. It simulates random touches, clicks, and other user events to make sure that abnormal usage won't make your app explode. The Monkey can be used to test your apps either on your emulator or on your own device. For more information, see `http://developer.android.com/guide/developing/tools/monkey.html`.

Paint.NET and GIMP

You will be working with images at some point in your Android development career. Most professionals use Adobe Photoshop, but not all of us can shell out that much money for an image-editing program. Therefore, you have two free alternatives: Paint.NET and GIMP.

Paint.NET is a free image-manipulation program written on top of the .NET Framework. Paint.NET works great and is used by many developers around the world. This application is targeted for Windows. Get Paint.NET here: `www.getpaint.net`.

The GIMP application is an open-source program that is similar to Photoshop. GIMP can be installed on Windows, Linux, or the Mac. See `www.gimp.org`.

Index

• *B* •

• *C* •

• J •

• K •

Internet

Blogging For Dummies,
3rd Edition
978-0-470-61996-4

eBay For Dummies,
6th Edition
978-0-470-49741-8

Facebook For Dummies,
3rd Edition
978-0-470-87804-0

Web Marketing
For Dummies,
2nd Edition
978-0-470-37181-7

WordPress
For Dummies,
3rd Edition
978-0-470-59274-8

Language & Foreign Language

French For Dummies
978-0-7645-5193-2

Italian Phrases
For Dummies
978-0-7645-7203-6

Spanish For Dummies,
2nd Edition
978-0-470-87855-2

Spanish
For Dummies,
Audio Set
978-0-470-09585-0

Math & Science

Algebra I
For Dummies,
2nd Edition
978-0-470-55964-2

Biology For Dummies,
2nd Edition
978-0-470-59875-7

Calculus For Dummies
978-0-7645-2498-1

Chemistry For Dummies
978-0-7645-5430-8

Microsoft Office

Excel 2010 For Dummies
978-0-470-48953-6

Office 2010 All-in-One
For Dummies
978-0-470-49748-7

Office 2010 For Dummies,
Book + DVD Bundle
978-0-470-62698-6

Word 2010 For Dummies
978-0-470-48772-3

Music

Guitar For Dummies,
2nd Edition
978-0-7645-9904-0

iPod & iTunes For
Dummies, 8th Edition
978-0-470-87871-2

Piano Exercises
For Dummies
978-0-470-38765-8

Parenting & Education

Parenting For Dummies,
2nd Edition
978-0-7645-5418-6

Type 1 Diabetes
For Dummies
978-0-470-17811-9

Pets

Cats For Dummies,
2nd Edition
978-0-7645-5275-5

Dog Training For Dummies,
3rd Edition
978-0-470-60029-0

Puppies For Dummies,
2nd Edition
978-0-470-03717-1

Religion & Inspiration

The Bible For Dummies
978-0-7645-5296-0

Catholicism For Dummies
978-0-7645-5391-2

Women in the Bible
For Dummies
978-0-7645-8475-6

Self-Help & Relationship

Anger Management
For Dummies
978-0-470-03715-7

Overcoming Anxiety
For Dummies,
2nd Edition
978-0-470-57441-6

Sports

Baseball
For Dummies,
3rd Edition
978-0-7645-7537-2

Basketball
For Dummies,
2nd Edition
978-0-7645-5248-9

Golf For Dummies,
3rd Edition
978-0-471-76871-5

Web Development

Web Design
All-in-One
For Dummies
978-0-470-41796-6

Web Sites
Do-It-Yourself
For Dummies,
2nd Edition
978-0-470-56520-9

Windows 7

Windows 7
For Dummies
978-0-470-49743-2

Windows 7
For Dummies,
Book + DVD Bundle
978-0-470-52398-8

Windows 7 All-in-One
For Dummies
978-0-470-48763-1

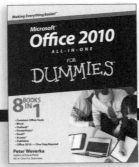

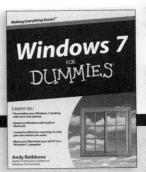

DUMMIES.COM®

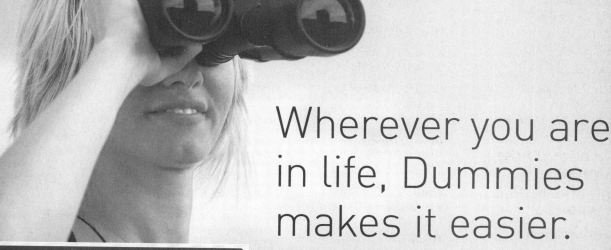

Wherever you are in life, Dummies makes it easier.

From fashion to Facebook®,
wine to Windows®, and everything in between,
Dummies makes it easier.

Visit us at Dummies.com

Dummies product make life easier!

DIY • Consumer Electronics • Crafts • Software • Cookware • Hobbies • Videos • Music • Games • and More!

For more information, go to **Dummies.com®** and search the store by category.

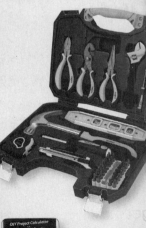